Zelalem Attlee

Polygonal Prostitution in Urban Ethiopia

Zelalem Attlee

Polygonal Prostitution in Urban Ethiopia

LAP LAMBERT Academic Publishing

Impressum / Imprint
Bibliografische Information der Deutschen Nationalbibliothek: Die Deutsche Nationalbibliothek verzeichnet diese Publikation in der Deutschen Nationalbibliografie; detaillierte bibliografische Daten sind im Internet über http://dnb.d-nb.de abrufbar.
Alle in diesem Buch genannten Marken und Produktnamen unterliegen warenzeichen-, marken- oder patentrechtlichem Schutz bzw. sind Warenzeichen oder eingetragene Warenzeichen der jeweiligen Inhaber. Die Wiedergabe von Marken, Produktnamen, Gebrauchsnamen, Handelsnamen, Warenbezeichnungen u.s.w. in diesem Werk berechtigt auch ohne besondere Kennzeichnung nicht zu der Annahme, dass solche Namen im Sinne der Warenzeichen- und Markenschutzgesetzgebung als frei zu betrachten wären und daher von jedermann benutzt werden dürften.

Bibliographic information published by the Deutsche Nationalbibliothek: The Deutsche Nationalbibliothek lists this publication in the Deutsche Nationalbibliografie; detailed bibliographic data are available in the Internet at http://dnb.d-nb.de.
Any brand names and product names mentioned in this book are subject to trademark, brand or patent protection and are trademarks or registered trademarks of their respective holders. The use of brand names, product names, common names, trade names, product descriptions etc. even without a particular marking in this work is in no way to be construed to mean that such names may be regarded as unrestricted in respect of trademark and brand protection legislation and could thus be used by anyone.

Coverbild / Cover image: www.ingimage.com

Verlag / Publisher:
LAP LAMBERT Academic Publishing
ist ein Imprint der / is a trademark of
OmniScriptum GmbH & Co. KG
Heinrich-Böcking-Str. 6-8, 66121 Saarbrücken, Deutschland / Germany
Email: info@lap-publishing.com

Herstellung: siehe letzte Seite /
Printed at: see last page
ISBN: 978-3-659-75549-1

Copyright © 2015 OmniScriptum GmbH & Co. KG
Alle Rechte vorbehalten. / All rights reserved. Saarbrücken 2015

Polygonal Prostitution in Urban Ethiopia. The Socioeconomic, Political Drivers and Public Health Intervention Imperatives.

Zelalem Attlee, MD, MHCM, DrPH(c)
International health policy study
2015

Contents

I. Definition and background..5
II. Analysis and results.........................…..15
III. Poverty and Human Development..30
IV. Urbanization and Slum Areas Expansion Compounding the Issue...36
V. Politics..44
VI. Legalization, Decriminalization, and Challenges.......................48
VII. Addressing Vulnerability and Recommendations.......................57
VIII. References...…..65

Preface

The issue of prostitution is important in developing countries nowadays because it is one of the key social issue that has multiple faces that impacts the general sexual reproductive health the population. This study was carried out between September 2005 and august 2010 in five major cities of Ethiopia i.e. Addis Ababa, Awassa, Bahrdar, Diredawa, and Nazereth. The author directly interviewed samples of women and men in each city and town who at that point in time were working in one of the mentioned types of prostitution. An extensive open-ended questionnaire was used and individual and focus group discussions were carried out which on average lasted 45 minutes. This is a prelude to upcoming social researches on human trafficking, gender studies, reproductive health, and related socio-epidemiological studies that will be carried out in the horn of Africa region. The case of Ethiopia is interesting because of its conservative culture and tight politics, which eclipse the public from these potentially detrimental social issues. Enjoy your reading.

I. Definition and back ground

Merriam Webster defines prostitution as 'the act or practice of engaging in promiscuous sexual relations especially for money (prostitution, 2014). Prostitution has been the oldest social institution in the world and its history in Ethiopia is not different from elsewhere. Different terminologies have been emerging in the past 100 years on depending on the type and extent of prostitution observed. Around the world prostitution is a billion dollar industry that is pursued mostly by the criminal underworld. Because of its nature and how its run prostitution has aged and matured to that level of being accepted as one way of life around the globe. However, the rights movements in the 21st century don't look at prostitution in that frame of mind. The very nature of purchasing women and children for sexual pleasure puts this occupation in the category of transgressions of basic human rights. It puts the victims in the domain of sub-humans going to the level of animals as far as human rights is concerned.

A similar term that is used by the NGO community is commercial sex, which is one of the similar connotations of prostitution but more elaborate than the gross definition. In fact commercialized sex is considered as a form of sexual exploitation and thus it is defined differently while maintaining the money transaction within its context. It is like seducing somebody to sell and use drugs in schools or to the general public at large without the will of the individual. So one end of the prostitution spectrum involves traffickers and pimps that feed on individual weaknesses be it financial, family issues, or the general gaps in the socioeconomic system per se; and on the other end prostitution eventually becomes coercive, deceptive, an act of abduction, and exploitative by force which displays victims as commodities in the world of commercial sexual exploitation. Because it is so commodified it becomes a market that differentiates itself into different shapes and forms for clients that utilize it. It ramifies itself into a huge market depending on the context of its geography controlled by the pimps or the traffickers in the form of frank prostitution, trafficking or pornography or it is served through different outlets as street-based sex work, erotic dance, stripping, cyber pornography, massage, saunas, and sex tourism. At the same time prostitution becomes a way of life mediated through poor choices in life. That further exposes the perpetrating individual back to the hands of pimps and traffickers and the cycle continues.

In the United States commercial sex is legally defined as 'The term "commercial sex act" means any sex act on account of which anything of value is given to or received by any person.' Be it legal or economical prostitution has a bottom line of combinations of abuse, which can physical, psychological or both, and it's a money transaction that involves human beings male, female, adult or child. In Ethiopia prostitution is legal, while pimping and brothel prostitution are illegal (100 Countries and Their Prostitution Policies, 2009).

The other face of prostitution is the domain of transactional sex. By definition, transactional sexual relationships are sexual relationships where the giving of gifts or services is an important factor. Transactional sexual relationships are distinct from prostitution, in that the exchange of gifts for sex includes a broader set of (usually non-marital) obligations that do not necessarily involve a predetermined payment or gift, but where there is a definite motivation to benefit materially from the sexual exchange (Hunter 2002). Often the participants frame themselves not in terms of prostitutes/clients, but rather as girlfriends/boyfriends, or sugar babies/sugar daddies (Hoefinger 2010, 2013). Those offering sex may or may not feel affection for their partners.

On the same plane we have the other dimension of prostitution that is known as survival sex. Survival sex is a form of prostitution, engaged in by people in extreme need. It describes the practice of people who are homeless or otherwise disadvantaged in society, trading sex for food, a place to sleep, or other basic needs, or for drugs (Barri, 2010). This again can be contextualized as a form of prostitution but entirely linked with economical tentacles. The multifaceted character of prostitution, and its behavior of duplication, interrelation with other faces of sex work and social determiners makes it more complex, subtle and polygonal.

Polygonal prostitution is the consummate result of external and personal factors that results in the different types of prostitution and finally becomes a generally accepted occupation in the social system. Polygonal prostitution results from a socioeconomic system that is repressive, where there is human rights abuse, continuing vicious cycle of poverty, high level of government corruption, poor infrastructure development in urban areas, weak policy implementation, and poor human development. Polygonal prostitution is the final leg of self-hate and loss of self-esteem, that becomes an accepted way of life irrespective of knowledge and education. In this book I will try to put on the table the factors that contribute for the emergence of polygonal prostitution and analyze the Ethiopian perspective in the synthesis of polygonal prostitution, and the factors that are attached to its sustenance.

This qualitative research additionally utilized the works of Harcourt - Donnovan framework when classifying the different types of prostitution, which detailed and classified at least 25 types of sex work were identified according to worksite, principal mode of soliciting clients, or sexual practices (Harcourt, Donovan, 2005). I used their framework to understand and further structure the types of prostitution that prevailed in the urban areas of Ethiopia. The study does not include rural areas, so that will require another framework and research. Just like the newly growing economies Ethiopia has a complex tradition, values and belief systems that reinforce sex work and expand those who are engaged in sex work alone or mixing it as an adjuvant job to beat poverty. The two scientists unveiled how prostitution is a multilayered social entity, which can seep deep into the socio cultural fabric of communities. The study has further coined an interesting dimension of how culture, and poor governance contribute to accentuate poverty, and give rise to the expansion of inner city unhealthy lifestyles like sex work and prostitution. This in turn clouds the cut off or demarcation point where regular normal living ends and most at risk groups engaged in poor sexual behaviors or most at risk groups who are engaged in sex work as an occupation starts.

Methodology

This study was carried out between September 2005 and august 2010 in five major cities of Ethiopia i.e. Addis Ababa, Awassa, Bahrdar, Diredawa, and Nazereth. The author directly interviewed samples of women and men in each city and town who at that point in time were working in one of the mentioned types of prostitution. An extensive open-ended questionnaire was used and individual and focus group discussions were carried out which on average lasted 45 minutes. The author located women and men engaged in prostitution through paid locator individuals who know the survey cities and towns very well. Interviews took place in restaurants and cafeterias (19.4%), on the street (20.2%), in clubs (25.7%), in respondent residences (16.5%), and in surveyor office (18.2%). Additionally the study group approached sex workers as clients and paid them for interviews only and not for sex services. A total of 950 women and 50 men over age 18 in the five study cities were interviewed. The author doesn't claim that this as a representative sample because of the lack of census material in this subpopulation and the illegal nature of prostitution in Ethiopia. On the other hand it is also very difficult to know and clearly demarcate the boundaries of this subpopulation as it takes new forms every time which in turn makes it difficult to clench what a representative sample should look like. On the other hand the author also believes that the interviewed sample is good enough to produce valuable information on the different issues related to sex work, redefining most at risk groups, ethical and psychosocial issues; and economic issues that drive the growth of this sub-population. The age range of interview participants was from 18 to 55.

Crude Results

Age at initiation

One of the key areas about prostitution is the age at initiation. The age at initiation correlates the household parent-child dynamics, school drop outs, socioeconomic status, and can predict future trends. From the total men surveyed engaged in commercial sex, majority 36% started between age 16-20 years, and 30% between 12-15 years. 2% of the interviewed men started below age 12 and were victims of childhood rape and molestation by a family member.

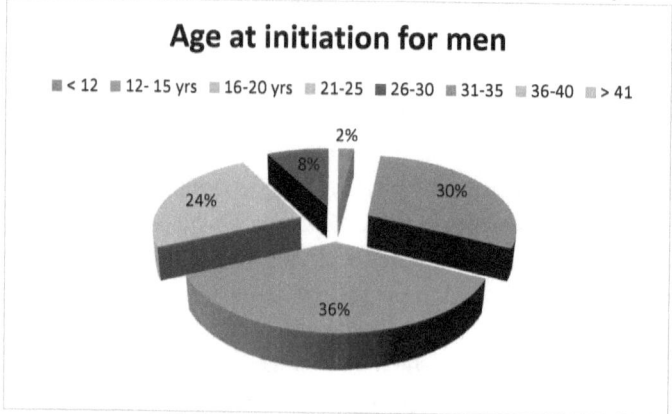

Figure 1 Age at initiation for men

On the other hand women have a different age pattern at initiation. Majority 22.3% of women were initiated between 21-25 years and 20.7% between 16-20 years. On the other hand women continue to enter into the commercial sex business irrespective of their age where 17% entered at 26-30 years, 16% at 31-35 and 12% between 36-40 years. Around 4% of the interviewed women were initiated below age 15 years and 8% of women started commercial sex after age 40.

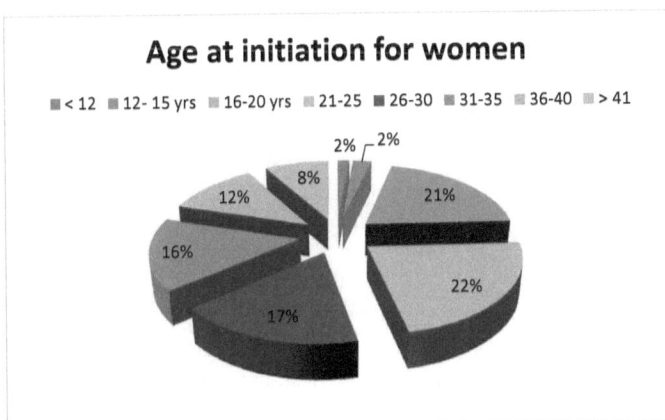

Figure 2 Age at initiation for women

Family dynamics

The majority of interviewed men (52%) have their parents still married but they live outside their home and 48% of men lived in *household* where there was single parent. On the other hand 55.7% of the interviewed women lived with a single parent, and left home either at early age or later on and 44.2% have come from homes while parents were living together.

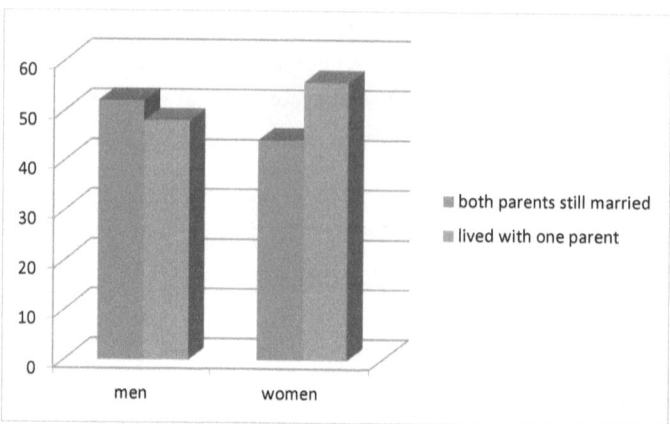

Figure 3 Family dynamics

Drug and alcohol use

The majority of men use khat and alcohol (96% and 100%) respectively. Only 12 % of the interviewed men used drugs mostly Marijuana. The khat and alcohol use pattern among the surveyed women has the same pattern (95% and 100%) respectively. Approximately 17% of the interviewed women additionally used Marijuana. From the interviews all participants both men and women used khat and alcohol on a daily basis and few of them added marijuana at least twice a week.

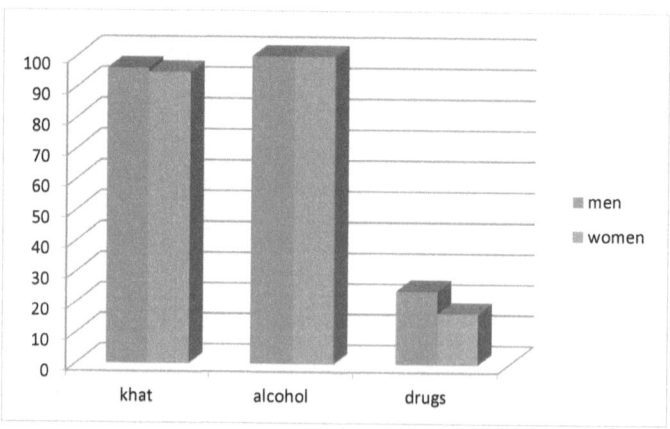

Figure 4 Drugs and Alcohol Use

Violence

The pattern of violence has different faces on the male and female sex workers. Male sex workers are highly marginalized and treated as low lives' in their community. Most community members whenever they get a chance tried to beat them up or insulted them, according to the surveyed men 96% have been attacked at least once in the last 12 months because of their profession and sexual orientation. However the violence from clients, sexual partners and at home was minimal at 24%, 28% and 36% respectively.

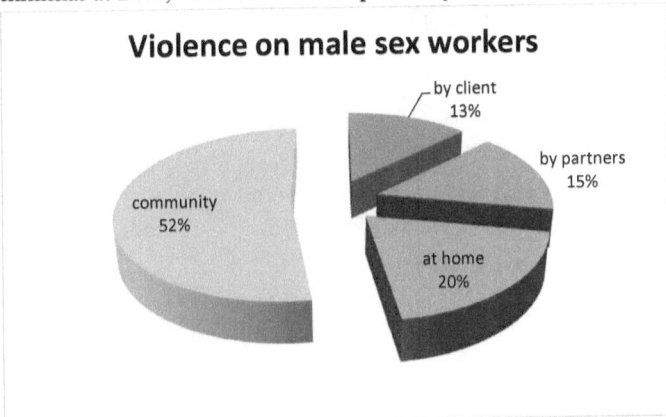

Figure 5 Violence on male sex workers

Violence on female sex workers came from clients, partners, and their household at 96%, 97%, and 99.7% respectively. Violence from community members where these sex worker females lived in was minimal at 2.6% showing the level of indifference to what they do irrespective of the knowledge about their profession by the community members.

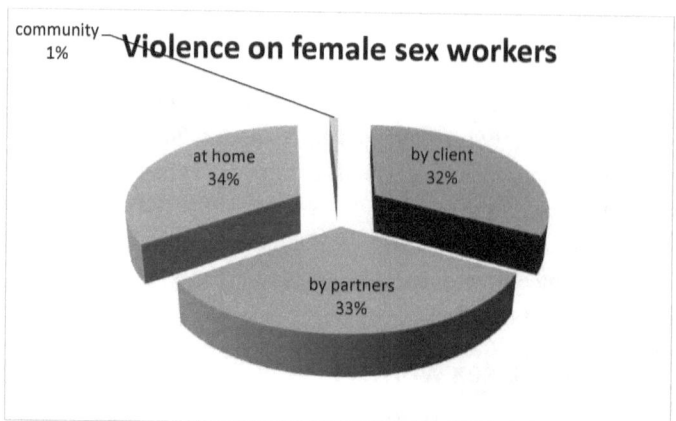

Figure 6 Violence on female sex workers

Workstation

70% of the men used clubs, cafeteria and hotels as their main contact points and work stations. Street and khat houses are used by 20% and 10% of male sex workers respectively.

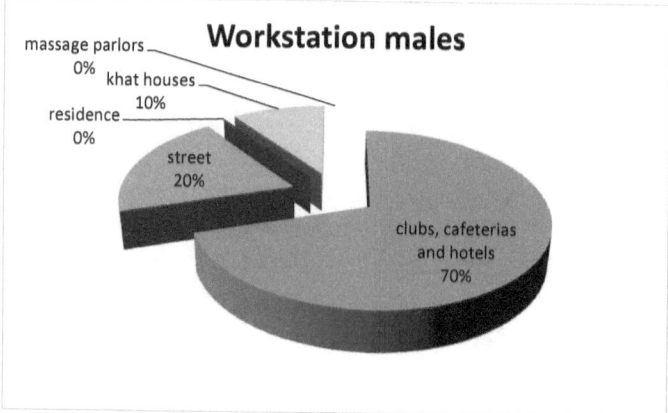

Figure 7 workstation for male sex workers

The majority of the interviewed women sex workers used the street as their main workstation 33.6 %, while clubs and khat houses are the next common workstations at 29.4 % and 21% respectively. Additionally massage parlors (10%) and calls from their residence (5.2%) were the key workstations for the interviewed female sex workers.

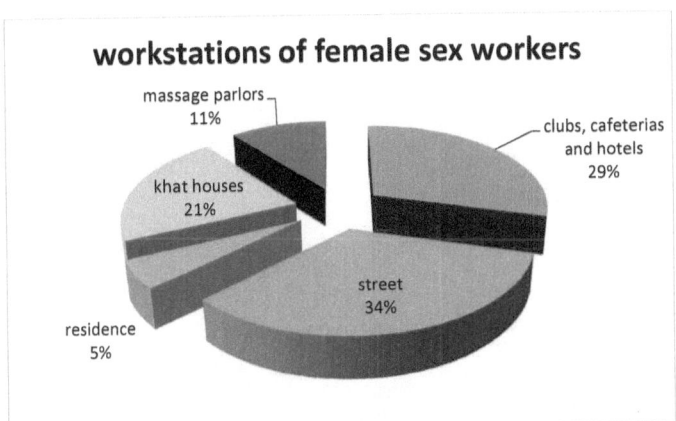

Figure 8 Workstation of female sex workers

Health issues

94% of the men and 64.4% of the women respondents disclosed that they had sexually transmitted illness at least once in the last 12 months. The diagnoses had been given by a medical doctor working in either a clinic or a hospital. All respondents refused to disclose their HIV status. 2% of the men and 36% of the women were homeless at the time of the interview.

Education

34% of the interviewed men have education level of 8th grade and below, 30% 12th grade level, 24% dropped out from college; and 12% were either in college or completed their first degree.

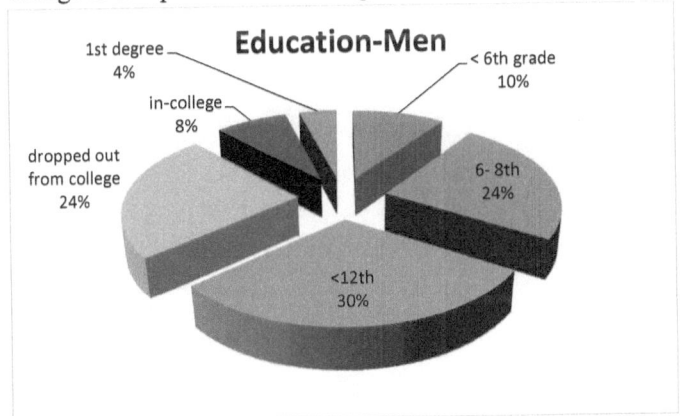

Figure 9 Education level for men

18% of the interviewed women had either 8th grade or lower level of education; 44% either completed high school or have dropped out from high school; 32% were either in college or have dropped out; and 6% had first degree.

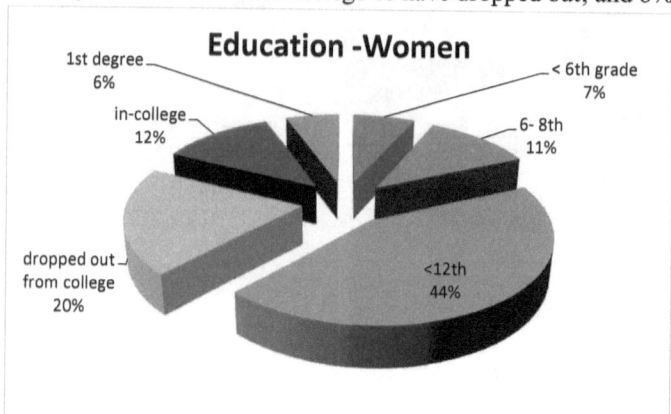

Figure 10 Education level for women

Looking at the overall comparative picture of educational status between the two genders most women who were high school dropouts or completes were involved in sex work, while higher percentage of men who had elementary school education level and college dropouts engaged in sex work. At the college level education women dominated the sex work industry.

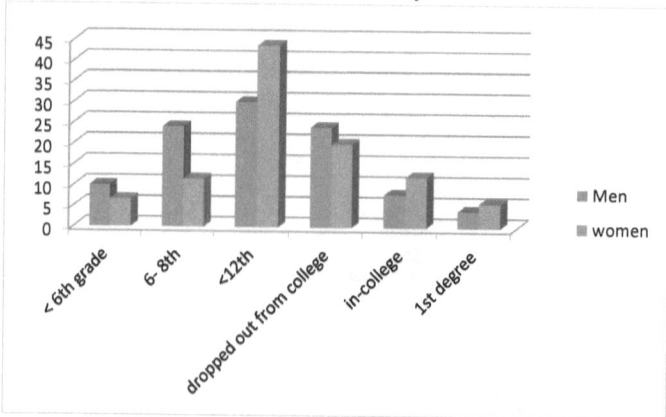

Figure 11 Gender comparison of sex work by educational level

II. Analysis of Results

One of the key areas about prostitution is the age at initiation. The age at initiation correlates the household parent-child dynamics, school drop outs, socioeconomic status, and can predict future trends. From the total men surveyed engaged in commercial sex, majority 36% started between age 16-20 years, and 30% between 12-15 years. 2% of the interviewed men started below age 12 and were victims of childhood rape and molestation by a family member. On the other hand women have a different age pattern at initiation. Majority 22.3% of women were initiated between 21-25 years and 20.7% between 16-20 years. On the other hand women continue to enter into the commercial sex business irrespective of their age where 17% entered at 26-30 years, 16% at 31-35 and 12% between 36-40 years. Around 4% of the interviewed women were initiated below age 15 years and 8% of women started commercial sex after age 40. The age of sexual initiation is associated with lack of employment, lack of comprehensive knowledge on HIV, alcohol and Khat use (Tilahun,Ayele, 2013). To the most part older adults play significant role in an early sexual initiation (Ompad et.al, 2006). The unique issue here when comparing the case of Ethiopian urban prostitution in females is that to the most part Ethiopian women start sexual intercourse in their late teens. The most difficult part in this study is the search and interview of male sex workers, where most of them have mixed economic background, and a lesser portion of them, 2% were victims of childhood rape by an older male adult. James, (2004) in his study states the following:
(1) The evidence for the hypothesis of maternal immune-reactivity to the male fetus is weak; (2) the evidence for the intrauterine hormone exposure hypothesis is also weak; (3) the evidence for the hypothesis of postnatal learning is stronger. Lastly, there seem likely to be causes common to male homosexuality and pedophilia. They may include sexual (or quasi-sexual) experience in childhood or adolescence (p.51-62).

The sexual experiences of male clients differ in the perception of the first initiation as an abuse or a normal initiation. In their study Carballo-diéguez, Balan, Dolezal, & Mello (2012) assessed the prevalence of recalled childhood sexual experiences with an older partner among men who have sex with men (MSM) and/or male-to-female transgender persons recruited, and associations between such recalled experiences and sexual risk behavior in adulthood in Campinas, Brazil. They state the following:

Participants' recruited using respondent-driven sampling completed a self-administered, computer-based questionnaire, and underwent HIV testing. For data analysis, raw scores were weighted based on participants' reported network size. Of 575 participants (85% men and 15%transgender), 32% reported childhood sexual experiences with an older partner. Mean age at first experience was nine years, partners being, on average, 19 years old, and mostly men. Most frequent behaviors were partners exposing their genitals, mutual fondling, child masturbating partner, child performing oral sex on a partner, and child being anally penetrated. Only 29% of the participants who had had such childhood sexual experiences considered it abuse; 57% reported liking, 29% being indifferent and only 14% not liking the sexual experience at the time it happened. Transgender participants were significantly more likely to report such experiences and, compared with men, had less negative feelings about the experience at the time of the interview. No significant associations were found between sexual experiences in childhood and unprotected receptive or insertive anal intercourse in adulthood. Results highlight the importance of assessing participants' perception of abuse, regardless of researchers' pre-determined criteria to identify abuse. MSM and transgender people may experience childhood sexual experiences with older partners differently from other populations (e.g., heterosexuals), particularly in countries with different cultural norms concerning sexuality than those prevalent in Europe and the U.S (p. 363-76).

Regardless of the above facts the issue of child molestation is one gray area that is completely ignored in the Ethiopian society and legal system. In fact when I was working on orphans and vulnerable children legal protection, the legal provisions as well as the implementations of what we already have in child protection law were nonexistent. This has increased the vulnerability of small boys and girls to child predators within their *household*, at school, or in their neighborhood.

Family dynamics

The majority of interviewed men (52%) have their parents still married but they live outside their home and 48% of men lived in *household* where there was single parent. On the other hand 55.7% of the interviewed women lived with a single parent, and left home either at early age or later on and 44.2% have come from homes while parents were living together. The fact that 52% of the studied subjects have both parents in marriage shows that beside broken homes due to divorce or parental death prostitution needs to be approached from multiple angles. In a qualitative study carried out on women prostitutes in Pakistan multiple pushing factors like poverty, worse economic conditions, illness in family, debt, sex for enjoyment, peer association, family neglect, domestic clashes, drug addiction in husbands and in involuntarily, forced rape, sexual assault, early marriages, trafficking, deceived by family, deceived by lover were observed (Qayyum, Iqbal, Akhtar, Hayat, Janjua, & Tabassum, 2013). Deciding based on the marital status of the family as stable household that prevents prostitution is a weak argument looking at the above facts. In fact the majority of the households in the studied cities are very poor economically speaking.

Drug and alcohol use

The majority of men used khat and alcohol (96% and 100%) respectively. Only 12 % of the interviewed men used drugs mostly Marijuana. The khat and alcohol use pattern among the surveyed women has the same pattern (95% and 100%) respectively. Approximately 17% of the interviewed women additionally used Marijuana. From the interviews all participants both men and women used khat and alcohol on daily basis and few of them added marijuana at least twice a week. In Ethiopia the most popular mild narcotic is khat which is officially sold in the streets of almost anywhere in the country. On the other hand in cities like Addis Ababa, and Awassa in this study I have found out that there are Marijuana users and the availability is ample. Prostitutes use drugs for various reasons like to fund the drug addiction itself, or to increase confidence, control, and closeness to others and to decrease feelings of guilt and sexual distress (Young, Boyd, Hubbell, 2000). The drug use pattern in Ethiopia is different from that of the west. Although in this study we have been told by few of the interview participants that they have used heroin especially with their rich clients in Addis, we couldn't confirm it otherwise. However illicit drugs can have deleterious effect and positively reinforce prostitution. Potterat, et.al, (19989) state the following:

> Drug use was more commonly reported by prostitutes than comparisons (86% vs. 23%), as was non-consensual pre-pubertal sex (32% vs. 13%). Sexual- and drug-related milestones occurred in the same order in both groups, with drug use preceding sexual activity and injecting drug use

preceding prostitution. Ninety-four percent of prostitutes who injected drugs reported non-injectable drug use before prostitution, and 75% of prostitutes who injected drugs reported doing so before beginning prostitution. The age distributions at critical events were similar for prostitutes and comparison women who reported regular drug use. Comparison women who did not report regular drug use were in general older than both these groups at the time of early sexual experience and drug experimentation. However, the ordering of these events was the same. Within the prostitute cohort, ethnic groups differed in their age distributions at several critical events, but not in the order in which these events occurred (pp.333-340).

Additionally the mild narcotic, Khat's long-term use has been associated with schizophreniform psychotic illness, mania, and more rarely, depression and also has negative impact on health and socio-economic conditions (Wabel, 2011).

Violence

The pattern of violence has a different pattern on the male and female sex workers. Male sex workers are highly marginalized and treated as low in their community. Most community members whenever they get a chance tried to beat them up or insulted the, according to the surveyed men 96% have been attacked at least once in the last 12 months because of their profession and sexual orientation. However the violence from clients, sexual partners and at home was minimal at 24%, 28% and 36% respectively. Ethiopia is a conservative country where morals and values are deeply rooted in religious belief systems. In Ethiopia homophobia and discrimination are the two fundamental anti-homosexual social effects affecting both lesbians and gay men. In this study I tried to find trans-genders at the time I couldn't find one. On homophobia (Hovey, 2007) states the following:

> The term *homophobia* once referred to fear of men but later came to mean fear or hatred of homosexuals. As with *racism*, it is a negatively charged word. It was coined by the psychologist George Weinberg, who used it in his 1971 book *Society and the Healthy Homosexual*. A phobia is an irrational fear bordering on a disorder. Calling antigay bias irrational implies that hating or fearing homosexuals is not sensible, normal, or healthy. Some people who stress the bias aspect of homophobia prefer the term *heterosexism* or *heterosexist*, which refers to the privileging of heterosexuality over other kinds of sexual expression, including but not limited to homosexuality. Heterosexism also stresses the social aspect of

homophobia as a learned and culturally reinforced set of attitudes rather than an illness or an irrational fear. Homophobia most often is used to characterize two kinds of attitudes: self-hatred and antigay bias. Self-hatred and shame about one's homosexuality often are referred to as internalized homophobia, implying that a homosexual person has adopted socially negative attitudes about homosexuality and that those attitudes damage that person's self-esteem. Trying to pass as a heterosexual or to cure oneself of homosexuality is seen by many people as a sign of internalized homophobia. Internalized homophobia also can be more subtle; one can be an acknowledged gay man and still not be like other gay men, or can be an acknowledged lesbian who refuses to hire or promote other lesbians (p. 715-717).

To the most part the situation in Ethiopia is at the stage of homophobia. But one can also argue to the fact that there is a stark social discrimination of homosexuals. From our interviews the type of discrimination against homosexuals is more individual than institutional at this stage. Violence on male homosexuals have resulted in serious physical and psychological disabilities in the past in the interviewed persons.

Violence on female sex workers came from clients, partners, and their household at 96%, 97%, and 99.7% respectively. Violence from community members where these sex worker females lived in was minimal at 2.6% showing the level of indifference to what they do irrespective of the knowledge about their profession by the community members. The type of violence depended on the workstation of the prostitutes and their age. Prostitutes working outdoors most frequently reported being slapped, punched, or kicked, whereas prostitutes working indoors cited attempted rape (Church, Henderson, Barnard, & Hart, 2001). Because of the extreme vulnerability of these prostitutes to violence most of them has post-traumatic stress disorder (Valera, Sawyer, & Schiraldi, 2003). In one study the life expectancy of a prostitute who started sex work as an adult is 34 years in Illinois (10 Facts that Shaped the Ugly Truth Campaign, 2011). The FBI states that the average life expectancy of a child once in prostitution is 7 years due to homicide and HIV/AIDS (Trafficking Stats, 2014). More data is needed in the case of Ethiopia on prostitution mortality due to violence and homicide.

Workstation and Types of Prostitution

In their literature review of the types of prostitution Harcourt and Donovan in 2005 had found around 25 types of prostitutions that are practiced in different settings. Furthermore, in their study Harcourt and Donovan have also further sub-classified the types of prostitution as direct and indirect prostitution which is a very important phenomenon indicating the sex work modalities of the prostitutes and their work station areas.

In this study 70% of the men used clubs, cafeteria and hotels as their main contact points and work stations. Street and khat houses are use by 20% and 10% of male sex workers respectively. The majority of the interviewed women sex workers used the street as their main work station 33.6 %, while clubs and khat houses are the next common workstations at 29.4 % and 21% respectively. Additionally massage parlors (10%) and calls from their residence (5.2%) were the key work stations for the interviewed female sex workers. In this study the following types of prostitution were observed using the Harcourt-Donovan framework:

The Different Types of Prostitution in the Five Studied Areas

In my five years of qualitative survey in Ethiopia that was carried out from 2005 and 2010 in the major urban areas of Addis Ababa, Awassa, Bahrdar, Diredawa, and Nazereth. The survey covered clients that are continuously using services from prostitutes and sex workers that are engaged in sex work either openly or discretely. We can classify the types of prostitution by gender as female and male forms of prostitution. However geography, financial status, religion, unemployment, and factors related to human development are included in these categories. The classification is based on client sex work utilization outlets around these major urban areas in the country. For the sake of simplicity I have classified the types of prostitutions as adult (male and female) and child prostitution. At the end of the survey I have found the following different types of male and female prostitution.

Adult Female prostitution

Red light prostitution: Red light prostitution goes down in liquor houses 'mesheta betoch' which can be whiskey bet, tella bet or tej bet, or a draft house with hired bar ladies. The bar ladies are regular employees of the bar/liquor house, but the prostitution is not included in the employment agreement, which is usually not in written form. Services are provided either in the back rooms of the liquor house or nearby smaller hotel with 'albergo' rooms. Red light prostitution is one of the cheapest forms of female prostitution which is commonly seen in almost all towns big and small in Ethiopia.

Street girls and women: In this class all age groups of the female gender starting from very young girls as young as 9 year olds to women in their 60's are involved. Street women stood up all night on the road sides starting from 7 PM in the evening till the morning hours. Clients can be street walkers or men passing by cars. Services are provided either in the car by turning around dark corners, or 'rental garages', or nearby regular households. Street girls hang out in almost all corners of the city of Addis Ababa, but mostly around the financial district of bole and similar areas.

VIP service: These are selected usually fashionable women who are transactioned by a mediator who can be a club owner, or a person that works for the wealthy people. The Client provides specification of the girl they need to be serviced to the middle person and the middle person brings the claimed girls to the client for the service to the premises. Gross bill is paid to the middle person and the share will be paid to the girl after service. Type of sexual services have individual charges. The girls that are called for this service can be working girls who have a working profession during the day time.

Escort girls: In Ethiopian the term escort girl has a different dimension. Like the VIP service escort girls are brokered by standard pimps or they can be regulars at three to five star hotels including resort hotels out of town. In some of the hotels picture albums of the girls along with a short biography are kept at the reception area with a focal contact person working in the hotel. The pimps will contact the escort girl by phone, and charges are paid to the middle person and the girl at the first encounter before or after the service depending how it is negotiated. Services are expensive, and mostly are used by foreigners which can be tourists or high level officials. Services are provided at private homes or hotel rooms.

Massage parlors: Massage parlors are everywhere in Ethiopia. The workers in the massage houses are mostly young women between teen age to their late 30's. Massage services are coupled with sexual services. There is a specific type of seductive dressing by the massage providers like bikinis and thongs and the massage process is not typically of a physiotherapic type. The girls are paid at the spot for their services and services are negotiated between the client and the service provider.

Kima Bet: Kima bet or Kima house is a house that has several rooms for chewing the stimulant plant Khat. There are simple Kima houses where there are several stools only, where clients chew khat for a while and take contact girls and go somewhere else for sexual services, or they can be of medium to high class where there are female service providers that lobby and facilitate the khat chewing ceremony, by chipping leaves off from the khat and feeding the clients along with massage. These girls are usually hired by the owner of the Khat house for the sole purpose of providing a waitress service to clients. But at the same time they negotiate sex with clients and services are provided either in rooms in the premises or outside the Khat house.

Private: A friend or a client contacts the sex worker usually by phone. Private clients can be married house wives or professional single ladies in which they are considered to be ordinary working women. The private type is more expensive, and the sex workers usually show disparity in their wealth in the neighborhood. While they are simple kindergarten teachers which in most cases don't have cars, they are seen to drive expensive cars and have an upper middle class type of living. Services are provided in client premises.

The Countess: Countesses are friends of rich people. They move in big circles surrounding rich people, but provide dating type of sex work. In this complex type of prostitution transactions can go down in the form solid cash, gifts, or opening businesses for these high class women. So long as the cash pipeline is open these women provide services to these rich men. They also play games between these rich men and solicit multiple cash sourcing by 'dating' different men from within the circle. These circles are so numerous and go by one mob, or the girls can jump between mobs and make money. Gentle women can also have a marriage to run, and they do the sex work on the side making this type of prostitution more complex.

Private strippers: private strippers are paid only for stripping off their clothing and dance nude. Private strippers are cross cutting which can be contracted out from high schools to all kinds of organizations. However some of them negotiate for sexual services and are paid in cash or in kind.

Pole dancers: This type of prostitution is seen in limited places in Ethiopia. Ploe dancing is similar as in the west where nude girls dance by holding poles and slowly stripping their bikinis. In Ethiopia these girls negotiate for sexual services and are paid in cash for their services outside or in the premise of the club.

Café girls: Girls who work in pastry shops and cafes also negotiate for sex but subtly. Transactions and services happen during the day time outside the premises of the café. Cash is paid for services.

Fird betoch : In English, fird betoch means court houses. These are members only clubs opened by women who are close friends and are friends with rich people. This is one of the most expensive sexual services provided in Addis Ababa. They sell liquor illegally, at a very expensive price limiting the service to few rich clients. The premises are closed at all times and entrance is only through codes or face scan by private security guarding the premises.

Bercha Garage: Is specifically practiced in Dire Dawa, in kima bets that only serve those who have money. The owners are usually Somali usually from Somali land, and girls are mixed Ethiopian and Somali girls especially imported and deployed for this service. In bercha garage a form of seductive positioning of the client and the sex worker naked but covering them with the traditional Horn of Africa Garment The Shirit. The Client will be fed with Khat, massage and seductive chatting along with alcoholic drinks while sitting in the tented Shirit. Services are provided at the spot and are paid to the owner which distributes the share for the sex worker later. Bercha Garage inclines towards the brothel business.

IL-Awos and Bibitos: These are open air tea parlors where the sex worker which basically is a tea maker female who serves tea to men chewing Khat. IL-Awos are very common in Somali communities near the studied towns. Since prostitution in the Somali community is strictly discouraged due to religion, transactions are carried out covertly, at a different premise and during the night time. Cash is paid for services.

University students: These girls are contacted through pimps either from the campus or from their rental houses down town. The pimp will be paid to bring the girls either to clubs or parties in private premises. Services are provided at the spot, and either cash or payment in kind is paid by clients. It is highly rampant in towns where there are big colleges and universities.

Lesbian cafes: Lesbians have a covert network in big towns and cities around Ethiopia. They have particular hang outs like café houses, or small clubs. Networking is through word of mouth, the internet (e-mails and/or social media). Lesbianism is believed to be imported especially it became more visible after the fall of the Dergue in Ethiopia. Lesbian prostitutes especially gave service to foreigners and they are paid in cash.

Camp prostitution: the sex workers move from camp to camp where ever there is huge construction. Camp prostitution is highly rampant in all studied areas because of the vast road and building constructions in the suburbs of the studied towns. They provide services to the camp workers who can be foreigners or locals. Cash is paid in the premises which are shelter houses near by the camps.

Sex tourism: Prostitution in tourist frequented spots around the country have grown in the past decade. Tour operation agencies advertise the indigenous people who live in the remote parts of the country in a sexual way in web sites around European countries. Many European tourists flock to these sites and experience the indigenous sex traditions with the locals. However in recent times the locals that I interviewed have confirmed that they demand payment for sex services that they provided. The sex tourism in Ethiopia blew in urban areas like Addis Ababa due to the explicit advertisement of Ethiopian women by touring agencies in Europe, and due to the long line of international conferences, and the booming hotel industry. Delegates from foreign countries and from major cities flock to these conference towns for conferences and retreat meetings. For powerful delegates there is always that connect that provides women at the end of the business day as part of relaxation and unwinding for the served guests. Sex tourism services are also provided as an extension of the city tour operation to rural areas like the deep south areas of Hamer. There services are provided in a 'jungle' fashion at night on the side of camp fires and remote huts nearby.

Survival sex: Survival sex is practiced by the homeless in urban areas around the country. It is entirely related to food security and surviving harms by gangsters in the street. Services are provided in nearby corners, or premises.

Modeling agencies: Recruit girls that are lined up for the modeling business to rich people. Services are provided in private premises, and transaction can be money for sex or in-kind for sex.

Adult Male prostitution

Male prostitution just like female prostitution has its both service ends for both men and women. In Ethiopia, the rise of male prostitution took a significant attention in the past 10 years. Male prostitution at least for now is confined in the metropolitan city of Addis Ababa. Addis Ababa due to its relative liberal stance accommodates new controversies that are pretty much outside of the cultural domain. Although no man admits his job is sex work in Ethiopia, we have tried to go beneath the surface and found the following types.

Gigolos: Gigolos can be regular working individuals, or they are 'contact escorts' serving women who come from abroad for vacation or for business. The customers served contact them through a network of previous customers, by phone or online. The dichotomy of the gigolo is that he can be a social affiliate with customer/s and can go in relationship with them while getting paid either in the form of stipends, remittance, or in kind. In most cases he lives with the customer, or is 'on-call' 24/7 by the customer as needed. Gigolos are generally dressed well, they work out, and they are taught to be classy by the woman when they communicate in the customer's social circles. The usual occupation of gigolos is modeling, but they can assume any job.

Gay for pay: In Ethiopia the gay network is covert and underground due to the conservative culture, at least for now. A person who provides Gay for pay services is an individual who may or may not be gay, but is paid for services. Customers are gay men who contact the payees through social media and the gay network in major cities around the country.

Most of these types of prostitutions are carried out either as frank sex work in bars or on the road or under the iron curtain of different businesses, connects and social networks. This very fact makes Ethiopia's prostitution spectrum very complex, complicated, polygonal and difficult to draw lines on vulnerability. The current context of the Ethiopian conservative culture and religious views and poverty have also played their role in escalating and expanding the diversity of prostitution transactions. The early 90's ushered the end of socialism and brought in capitalist oriented governance in Ethiopia after the overthrow of the Dergue by the EPRDF. The change in the economic system resulted in the cropping up of new multi-millionaires and oligarch elite in the system putting out new demands in the prostitution market. Coupled with that the high influx of tourists, conference delegates, the booming hotel industry and Ethiopian diaspora have contributed to the diversification of the prostitution business in Ethiopia. Ethiopian girls are openly advertised in international tourist web sites as the most beautiful and highly sought commodities in the business. This has brought in a web of pimping networks and connections to deliver the services to foreign and domestic clients seeking the services. The previous conservative generations have aged out, and a new business hungry generation is bearing the brunt of sex work and is taken advantage of by rich people, tourists and pimps. The interesting finding in this study is that the involvement of educated women and men as gigolos in the prostitution business. The reason behind it is not poverty but 'low' incomes which the sex work will fill the gap to get the fancy elements like expensive clothing, cars and make ups. It indicates that prostitution can also be a matter of choice rather than being a bitter opportunity/possibility. The intense bling of the hip hop influence has sipped so deep in the modern urban Ethiopian society; the life style of the hip hop stars on the video has created so much craving in the young generation. So creating wealth through illegal short cuts is the current way of getting money, and getting out of poverty. This has masked the value of education, morals and ethics in the educated and the elite, and getting money through such sex work transactions is considered normal and everybody has blended in it. Back in the day everyone worked hard to earn their promotion in their job and careers and sweated to earn what they have achieved in their life time. Now that is completely gone. To the most part every person in the modern urban Ethiopian society is continuously creating illegal schemes of sponging money out of the system, and out of individuals who got it. This doesn't mean though there are no honest hard working people in Ethiopia. It doesn't rise to that level of stereo typing. However there is a clear trend of a Ponzi generation in the making, and all the above types of prostitutions are but one face of that tip of the ice berg.

Inter-Generational sexual relationships

The growing wealth gap in Ethiopia is contributing to divergent sexual relationships between couples especially in age. While in Ethiopian history there have been such relationships manifested as forced marriages between an old man and a young girl to maintain family wealth and power in the rural parts of the country. At present while there are no officially announced 'forced marriages' in Ethiopia , due to gaps and disparities in socioeconomic factors like wealth and power, cross generational sexual relationships have become rampant in most parts of the country. In the studied capital towns and metropolitan cities this has been true because of the clearly visible socioeconomic and gender disparity. In Ethiopia working men will have well established business and wealth by the time they reach well into their forties and fifties. This attracts younger girls to quicken their pace up in life's ladder and they will start dating older men to fulfill their needs. The needs of the girls range from the basics to higher career and/or business related, which will be fulfilled by these old men. Because of this fact most young girls prefer this short cut instead of advancing themselves step wise through education or hard work. This has set a culture of dependency of both younger men and women in most societies around the country. The same holds true for younger boys. Although not as rampant as the girls, this author has encountered many of such scenarios especially in Addis Ababa. Single older women who came from abroad with stacks of dollars and wealth have younger boys to their choice of sexual pleasure keeping them as sexual partners and supporting them financially to achieve their objectives. This had also been part of the culture in previous times, where a married infidel woman would keep a boyfriend somewhere in town to satisfy her bodily needs. Married couples are not exempt from this cross generational sexual relationships, especially those couples who have lived together for decades and accumulated significant wealth, and have worn out passion for each other. How many of them are like that? Well we can say countless. This paradigm of cross generational sex has been blamed to be the prime factor in transmitting HIV and STDs , gender based violence, unwanted pregnancies; and abortion. The rise in sugar daddies and sugar mummies has contributed to the diversification of prostitution and transactional sex in Ethiopia. In such kind of sexual relationship the commercialization of the sex is not usually in liquid money alone, it comes combined with gifts starting from smaller items to larger items like houses or expensive cars. To the most part older men or older women if they have the wealth they will make an imposition on their younger partners not to use condoms after the first few 'test runs' and convince them to engage in sexual acts without protection. This ultimately will lead to higher risks of HIV . (Gregson et.al, 2002) states the following:

> Older age of sexual partner was associated with increased risk of HIV-1 infection in men (odds ratio 1.13 [95% CI 1.02-1.25]) and women (1.04 [1.01-1.07]). Young women form partnerships with men 5-10 years older than themselves, whereas young men have relationships with women of a similar age or slightly younger. Greater number of lifetime partners is also associated with increased risk of HIV (1.03 [1.00-1.05]). Young men report more partners than do women but infrequent coital acts and greater use of condoms. These behavior patterns are underpinned by cultural factors including the expectation that women should marry earlier than men. A strong gender effect remains after factors that affect exposure to infected partners are controlled for (6.04 [1.49-24.47]) (p. 1896-903).

The author in his interviews with research subjects has encountered a wealthy person bragging that he has 6 kept women in 6 villas and apartments across the country. Discordant partnerships in terms of race/ethnicity, age, education, and number of partners have been shown to increase the risk for STDs (Aral et.al, 1999). In this study while socio economic disparity between older wealthy men and younger girls is the driving factor for the inter-generational sex and all the risks associated with it, we cannot blame poverty as a sole factor for its existence. Poor decision making on the part of younger girls, and hunting for flashy life through gold digging path is equally contributing for this intergenerational sexual relationship. Almost a quarter of the interviewed young girls in universities, and those who have professional careers don't live in poverty. 30% of the research participants have a permanent sexual partner as a boyfriend, co-habitant or a husband. However one key observation from this qualitative research is the rise in extra dyadic sex to astronomical proportions. In these extra dyadic relationships women are more attached to their sexual partner as compared to men, and the age, number of sex partners, number of years married/cohabiting and sexual orientation are key factors related to extra dyadic relationships (Træen et.al, 2007).

Inter-generational sex is the key gray area in the commoditization of sex, because it exists below the surface and is practiced across all levels of society irrespective of ethnicity, socioeconomic status, age, and sexual orientation. Its occurrence indicates a high level of promiscuity within the social system opening ways to polygamous, polygynous and perhaps polyandrous sexual relationships. The Ethiopian culture to its most part is not open about discussing sexual relationship issues except in the metropolitan areas. Prostitution is legal in Ethiopia. However, the lack of openness in the society coupled with greed, poverty, and 'dream civilization' has led to the lack of self-esteem, self-respect which led to believing and justifying such practices are alright in individual discussions. This also has led to the high rate of human trafficking of men, women and children under the label of economic migration while the truth is human trafficking for sex (especially women and children) to the Middle East, Eastern Europe and south East Asia.

Health issues

In this study 94% of the men and 64.4% of the women respondents disclosed that they had sexually transmitted illness at least once in the last 12 months. The diagnoses had been given by a medical doctor working in either a clinic or a hospital. All respondents refused to disclose their HIV status. 2% of the men and 36% of the women were homeless at the time of the interview. Sex workers are vulnerable to HIV, STDs like syphilis, gonorrhea, hepatitis B and chlamydia (Resl, Kumpova,Cerna, Novak,& Pazdiora, 2003). Also psychiatric disorders like PTSD, depression, and anxiety are common in trafficked women and women exposed to violence (Hossain, Zimmerman, Abas, Light, Watts, 2010). Studies also show the direct correlation of substance abuse and prostitution. Ninety-four percent of prostitutes who injected drugs reported non-injectable drug use before prostitution, and 75% of prostitutes who injected drugs reported doing so before beginning prostitution (Potterat, Rothenburg, Muth, Darrow, & Phillips-Plummer, 1998).

Education
In this study 34% of the interviewed men have education level of 8th grade and below, 30% 12th grade level, 24% dropped out from college; and 12% were either in college or completed their first degree. 18% of the interviewed women had either 8th grade or lower level of education; 44% either completed high school or have dropped out from high school; 32% were either in college or have dropped out; and 6% had first degree. Looking at the overall comparative picture of educational status between the two genders most women who were high school dropouts or completes were involved in sex work, while higher percentage of men who had elementary school education level and college dropouts engaged in sex work. At the college level education women dominated the sex work industry. Educated women working as part time prostitutes has been reported in Iran (Iran's educated, middle class, par time prostitute, 2014). In another study that is carried in the University of Arkansas analyzed the US prostitution market stated "The findings suggest that these women are not forced into the prostitution market but rather choose to enter it for many of the same reasons that people enter the conventional job market – money, stability, autonomy and even job satisfaction (Affluent, Educated Women May Choose Sexual Prostitution, 2014)." The other interesting recent development was the porn star from Duke University, Belle Knox that stirred huge controversy on her remarks that states sexual liberation and women empowerment (Piers Morgan Interviews Pornstar Belle Knox Duke, 2014).

Having an economic growth in double digits, where everybody is somehow making money in most parts of urban Ethiopia, having the context of this age of the internet and women rights can one simply conclude that Ethiopia is another moving part in this machinery? Well that has yet to be answered in the coming decades, on how the issue of women empowerment evolves. For me women empowerment deals with the abatement of discrimination of women by gender, race, ethnicity, social class and religion. At the same time it works towards solving areas of inequality where women were marginalized.

The other spectrum of women empowerment is feminism. Although there are many types of feminism, the four most common are liberal that accepts that sex differences exist but contends that social, legal, and economic opportunities should be equal for men and women; radical feminism argues that men are oppressors of women and that the patriarchal social structure must be replaced for women to gain equality and it represents many divergent groups, including cultural feminism, lesbian feminism, and revolutionary feminism; Marxist feminists believe that women's oppression stems largely from economic stratification brought about by the production methods inherent in capitalism and capitalism must be destroyed in order to emancipate women both as workers and as property within the marital sphere; and socialist feminists argue that both class and sexism are sources of women's oppression they advocate the end of capitalist patriarchy to reduce all forms of exploitation, as they are also concerned with oppression resulting from race, age, religion, and the like (Feminism, 2008). So the Ethiopian context of education and prostitution can take one of the above feminist stances mixed with a cropping up attitude change amidst economic development and civilization. As far as liberation of women in Ethiopia is concerned Addis Ababa is by far advanced than the rest of the towns surveyed, because it is a melting pot of all Ethiopian cultures and traditions and it is more advanced as far as the main stream psyche of the population towards women is concerned.

III. Poverty and Human Development

One of the evil faces of inequality around the globe is poverty. Poverty depends on the overall socioeconomic context of the country in question. In developing countries It presents as an absolute entity where poverty is marred by starvation or relative as in developed countries where millions of families struggle to meet the basic costs of housing, food, transport, heating and health care (Sadler, Philip,

2010). Poverty is one of the social determiners around the world that has persisted for so long for many decades or in some countries' context centuries. The most poverty stricken region in the world is
south Asia and sub-Saharan Africa. The number of poor people in Africa doubled between 1981 and 2005 from 200 million to 380 million, and the depth of poverty is greater as well, with the average poor person
living on just 70 cents per day; with poverty rate unchanged at 50 per cent since 1981due to fast population growth ; where as in south Asia there are more people in absolute numbers with 595 million, of
which 455 million live in India but the poverty rate, however, has fallen from 60 per cent to 40 per cent (Sadler, Philip, 2010).
Ethiopia resides in the horn of Africa, a region with high rates of crisis in neighboring countries like Somalia, and south Sudan, and with one of the lowest urbanization. Ethiopia is the second populous country in sub-Saharan Africa next to Nigeria. According to the world bank in 2013 the estimated population of Ethiopia is 94.10 million, has a GDP of 46.87 billion, an inflation rate of 8.1%, has 29.6% of the population below the poverty line and GDP growth of 10.4% (The world bank, 2014). The definition of poverty varies according to the country in question.
Choosing and Estimating a Poverty Line (2011) states the following:

> Poverty lines are cut-off points separating the poor from the non-poor. They can be monetary (e.g. a certain level of consumption) or non-monetary (e.g. a certain level of literacy). The use of multiple lines can help in distinguishing different levels of poverty. There are two main ways of setting poverty lines—in a relative or absolute way: **Relative poverty lines:** These are defined in relation to the overall distribution of income or consumption in a country; for example, the poverty line could be set at 50 percent of the country's mean income or consumption. **Absolute poverty lines:** These are anchored in some absolute standard of what households should be able to count on in order to meet their basic needs. For monetary measures, these absolute poverty lines are often based on estimates of the cost of basic food needs (i.e., the cost a nutritional basket considered minimal for the healthy survival of a typical family), to which a provision is added for non-food needs. For developing countries, considering the fact that large shares of the population survive with the bare minimum or less, it

is often more relevant to rely on an absolute rather than a relative poverty line. Others include like **the food-energy intake method which** defines the poverty line by finding the consumption expenditures or income level at which a person's typical food energy intake is just sufficient to meet a predetermined food energy requirement. If applied to different regions within the same country, the underlying food consumption pattern of the population group just consuming the necessary nutrient amounts will vary. This method can thus yield differentials in poverty lines in excess of the cost-of-living differential facing the poor. **The Cost of Basic Needs method** which values an explicit bundle of foods typically consumed by the poor at local prices first. To this, a specific allowance for nonfood goods, consistent with spending by the poor, is added. However defined, poverty lines will always have a high arbitrary element; for example, the calorie threshold underlying both methods might be assumed to vary with age (p.1).

The picture of poverty is the same around the world. Just like any poor person in Africa a poor person in the United States can be totally homeless without any form of income or lives in a shabby house, in a dilapidated neighborhood. However due to the existing social infrastructure poverty in the US can be handled differently through different institutional mechanisms. Poverty in the US has its own income brackets based on family size, and it differs from state to state. Here is one table that shows poverty brackets for the District of Columbia, Alaska and Hawaii. These are called poverty thresholds and poverty guidelines, where the previous is the original measure updated by the Census bureau and the latter is the other measure issued each year in the *Federal Register* by the Department of Health and Human Services (HHS). According to the 2013 poverty guidelines, (2013), here are three poverty guidelines for three states issued by the federal register:

2013 POVERTY GUIDELINES FOR THE 48 CONTIGUOUS STATES AND THE DISTRICT OF COLUMBIA	2013 POVERTY GUIDELINES FOR ALASKA	2013 POVERTY GUIDELINES FOR HAWAII

		Persons in family/household	Poverty guideline	Persons in family/household	Poverty guideline
For families/households with more than 8 persons, add $4,020 for each additional person.		For families/households with more than 8 persons, add $5,030 for each additional person.		For families/households with more than 8 persons, add $4,620 for each additional person.	
1	$11,490	1	$14,350	1	$13,230
2	15,510	2	19,380	2	17,850
3	19,530	3	24,410	3	22,470
4	23,550	4	29,440	4	27,090
5	27,570	5	34,470	5	31,710
6	31,590	6	39,500	6	36,330
7	35,610	7	44,530	7	40,950
8	39,630	8	49,560	8	45,570

SOURCE: *Federal Register*, Vol. 78, No. 16, January 24, 2013, pp. 5182-5183

According to the above tables for example a single poor person in Hawaii earns at least 1102.50 USD a month. If we look at the definition of income poverty in developing countries it has a different dimension. According to these estimates, 21 percent of people in the developing world lived at or below $1.25 a day in 2005; in all, 2.4 billion people lived on less than US $2 a day in 2010, the average poverty line in developing countries and another common measurement of deep deprivation (The World bank, 2014). The situation of income poverty in Ethiopia is one of the contributing factors for the social problems which are becoming increasingly wide spread where prostitution is one of them. Poverty is not entirely related to low incomes only. It is a multi-faceted subject that has many faces and causes. The following factors can be included as parts of poverty: lack of security, poor gender relationships, abuse by those in power, problems in social relationships, excluded locations, limited capabilities, physical limitations, precarious livelihoods, dis-empowering institutions and weak community organizations(voices of the poor, 2012). This leads to the conclusion that the human factors play in making a personal choice to pursue the profession of prostitution, and in the mix there can be poverty as one pushing factors. Studies have further confirmed that the pushing factors towards prostitution are poverty, worse economic conditions, illness in family, debt, sex for enjoyment, peer association, family neglect, domestic clashes, drug addiction in husbands and in involuntarily, forced rap, sexual assault, early marriages, trafficking, deceived by family, and being deceived by lover (Qayyum et.al, 2013). From the above results of this study for example the fact that 100% of all women being initiated by a close girlfriend but not enforced by an external authority, and 24% men and 32% women were either in college or dropped out can imply that these sections of the society can engage in other regular works and can make a sound choice beside prostitution. The other more interesting fact that comes out from this study in the poverty argument is that when looking at the different types of female and male prostitutions most of the types were carried out in some combination. Especially those types of prostitution like VIP service, private, the countess, private strippers, modeling agencies, and gigolos are carried out on the side as second line income generators by the perpetrators again leading to a dichotomy in the thinking of poverty as a key driver or pushing factor to prostitution. Studies have also argued this fact by stating that prostitution is determined by 3 factors which are first, the moral and cultural pattern of the prone person, which is generated, above all, by the cultural, family, and relational environment (see the rare frequency of this practice in the educated, ultra-religious, environments, and in small communities where social control is more intense); second, the economic motivation required for the practice of this occupation (the sex industry is concentrated in the rich states); and to a lesser extent, the risks generated by the negative social reaction of communities where the practitioner works, the extent of social control and the tolerance toward such practice of the community members (for example, the social tolerance within numerous communities in Spain or Holland and partially in Germany) (Andreescu, Zaharie, 2014).

The human development index shows the level of human development on life expectancy, education and income. It's all about whether there is what they call a decent standard of living in a given country. These factors are also coined in the millennium development goals developed by the United Nations. (Debroy, 2006) states the following on HDI:

> HDI is the most commonly cited index. It is based on three variables: a long and healthy life (measured by life expectancy at birth); knowledge (measured by adult literacy rate and gross enrollment ratio); and a decent standard of living (measured by per capita gross domestic product in PPP U.S. dollars). The aggregation process is such that the maximum value of

HDI is 1 and the minimum value is 0. The higher the HDI value, the better. Depending on the HDI value, countries are divided into three categories: high human development for countries that have an HDI more than 0.800; medium human development for countries that have an HDI between 0.500 and 0.800; and low human development for countries that have an HDI less than 0.500 (p.215-219).

Debroy further filters the goals related to that in the MDGs relevant for the socioeconomic issues as follows:

Goal 1	**Target 1: Halve, between 1990 and 2015, the proportion of** people whose income is less than $1 a day
	Target 2: Halve, between 1990 and 2015, the proportion of people who suffer from hunger
Goal 2	Target 3: Ensure that, by 2015, children everywhere, boys and girls alike, will be able to complete a full course of primary schooling
Goal 3	Target 4: Eliminate gender disparity in primary and secondary education, preferably by 2005, and in all levels of education no later than 2015
Goal 4	Target 5: Reduce by two-thirds, between 1990 and 2015, the under-five mortality rate
Goal 5	Target 6: Reduce by three-quarters, between 1990 and 2015, the maternal mortality ratio
Goal 6	Target 7: Have halted by 2015 and begun to reverse the spread of HIV/AIDS
	Target 8: Have halted by 2015 and begun to reverse the incidence of malaria and other major diseases
Goal 7	Target 9: Integrate the principles of sustainable development into country policies and programs and reverse the loss of environmental resources
	Target 10: Halve, by 2015, the proportion of people without sustainable access to safe drinking water and basic sanitation
	Target 11: Have achieved by 2020 a significant improvement in the lives of at least 100 million slum dwellers

Source: Debroy, 2006

Ethiopia's HDI value puts the country in one of the lowest HDI categories. Ethiopia's HDI value for 2012 is 0.396—in the low human development category—positioning the country at 173 out of 187 countries and territories; between 2000 and 2012, Ethiopia's HDI value increased from 0.275 to 0.396, an increase of 44 percent or average annual increase of about 3.1 percent; in Ethiopia 87.3 percent of the population lived in multidimensional poverty (the MPI 'head count') while an additional 6.8 percent were vulnerable to multiple deprivations; the intensity of deprivation – that is, the average percentage of deprivation experienced by people living in multidimensional poverty – in Ethiopia was 64.6 percent (Human Development Report 2013). There is no doubt that poverty plays a major part in the diversification of prostitution in the country. On the other hand one can infer that prostitution is still negatively affects the life expectancy, schooling, HIV prevalence, and the overall perception of positive living in the affected and vulnerable population.

IV. <u>Urbanization and Slum Areas Expansion Compounding the Issue</u>

The economic inequality and disparity in *household* income in urban areas of developing countries is widening due to increased rural-urban migration, the expansion of slum areas, weak government policies and due to the creation of small concentric affluences amidst the poor. In Addis Ababa for example almost half of the city households reside in or around slum areas. These slum areas are expanding each year and the corrupt land grab practice pursued by the city council has further fueled its expansion to neighboring regions and zones. People in search of a better livelihood migrated from rural areas to major urban cities causing a massive explosion of urban population around the cities of the world. Rural-urban migrations can also result from the search of basic infrastructure like electricity, hospitals and safe drinking water (Eyo, Ogo, 2013). Studies show that by 2050 more than 70% of world population will live in cities (Urban Health, 2014). When people move from what normally was a well settled rural situation to new environment they generally start from scratch to catch up with the urban lifestyle. Because most of the migrants to these major urban cities start from the lowest economic level coupled with the ongoing urban poor population increase they end up living in the poor corners of these cities. These poor areas are commonly known as slums. Slum areas are characterized by abject or near poverty, poor infrastructure like housing, electricity, road development, safe drinking water and poor waste disposal systems. The urban population in developing countries is growing by 36% and by 2001 31.6% of the world population lived in slum areas, where Sub-Saharan Africa has the highest proportion at 71.9%, and by absolute numbers Asia has the highest number of slum dwellers at 554 million in 2001 which is about 60% of the world's slum dwellers(The challenge of Slums, 2003). Slum areas are characterized by poor health services access, and quality, social stigmatization of the dwellers, social disorganization, and social problems like prostitution, drugs, crimes and violence.

The high rate of urbanization in Ethiopia is directly linked with the formation of different social and economic classes within cities. This heterogeneity has its own ripple effects on the general public health of urban areas which is associated with overcrowding, poor transportation, poor sanitary facilities, and poor safe drinking water supply and sewage systems on top of other social problems like crime, violence and prostitution. The increased rate of rural-urban migration will bring in migrant workers from rural areas in to major cities. These poor migrant workers will double the population living in poverty and with poor access to health care services. This strains the overall health care financing to the population of the cities, further compounding the availability of basic health services like immunization, and preventive care. Furthermore the labile population in slum areas causes difficulties in public health data managements on pin pointing basic epidemiological data like morbidity and mortality data for diseases that exist within the slum areas.

Because slum areas expand with poor house planning, diseases that thrive in overcrowded situations will increase in prevalence like tuberculosis, communicable infectious diseases like upper respiratory tract illnesses, STDs, and diarrheal diseases. The stagnant economy in slum areas coupled with the lack of basic social services like schools, police security, and lack of implementation of government regulatory policies opens doors for illegal underworld trade like drugs, trafficking of women and children, crimes and violence which are associated with high rates of morbidities and mortalities. Poor health services that are associated with maternal and child health care are responsible for high mortalities in slum areas. High levels of maternal mortalities due eclampsia, post-partum hemorrhage and child mortalities like sepsis, birth asphyxia and birth trauma are seen in slums

(Khatun, et.al, 2012). Additionally high percentage of homicides, mortalities due to drug and alcohol intoxication especially due to boot legged brews are also common in slum areas (Ziraba, Kyobutungi, & Zulu, 2011). Slum areas are deprived of basic social services like policing, schools, and community centers to alleviate issues of security, and education.

Physical, Social, and Environmental Causes

Slum areas are formed due to rapid influx of rural population in urban areas. The unique feature of Ethiopia's slum areas are they are so mixed in the studied cities and there is no clear demarcation between the affluent sections and the poor sections in the studied urban areas. An estimated half a million people move in and out of Addis Ababa every day, leading to a staggering rate of rural-urban migration and poverty expansion. This rural-urban migration creates a situation of rapid urbanization of cities, and leads to expansion of slum areas due to the high influx of poor rural population to cities. The recent economic growth in Ethiopia has resulted in a shift of social relations between the different socioeconomic classes in the studied urban areas. The bottom line observed in this economic shift is that there is a high level of corruption in almost all sectors of the government, giving rise to 'accidental riches', and the degradation of civil structures through extreme bribery, and bending of government regulations. This gave way to an increased income inequality, and wealth gaps leading to formation of economic classes in these urban areas. The increased class segregation produced concentric affluent and poor groups all living together in many of the studied towns. Because the government's town planning methods wouldn't give way to affluent neighborhoods which are gated as in the western countries the town planning modalities of the Ethiopian government city planners look only for the availability of a spot to build a given condominium for those who have been displaced from one demolished slum area in the cities to the new spot. Because of this practice the poor and the rich basically live together resulting in concentration of affluence mixed amidst poor sections of the town. In developing countries this is one huge gray area where housing conditions are not addressed in slum areas, leaving slum areas in squalid living conditions, and poor housing. Studies show that in the developing countries of Latin America the slum typology shows that in the center city there are the hand-me-down housings, the squatters, the pavement dwellers, the informally rented poor houses, and the internally displaced people (Schrader, 2007). While the majority of the population is poor in the studied towns the density of poverty led to the rise of high crime, prostitution, family dissolution, disease and violence. In most of these studied towns the rich tend to get more privileges, and initiate new demands in the sex industry, creating more differentiation and expansion of the sex work market. Each rich person has a posse and entourage that would provide pimping and delivery services of young economically poor girls for transactional sex services for the rich. Pimping services can take the form of violent approaches that rise to the level of gang rapes by thugs that literally kidnap and put the girls in a room for displays or line-ups so that the rich person chooses the one he likes and force her to do sex. This mechanism basically reinforces the ecological disruption of these urban areas, and the class advantage of the concentrated rich on the poor compounding the deep class division and segregation of this violent social environment seen in these studied towns. In continents like Africa where there is poor economic growth the high influx of poorer people to urban areas creates a high labor supply which cannot be absorbed by the weak economies, which in turn creates high unemployment (Kempe, 1999). On the other hand the on-going industrialization and modernization of cities creates high level of competition for the current legitimate jobs available and when that saturates to the expansion of underworld economies of the slum, leading to slum expansion. Rural-urban migration is also caused by high yield agricultural practices which led to low crop prices, conflicts, international migration, and natural calamities pushing the younger work force to urban areas and leaving the older population that is on government well fare in rural areas (T, M. A., Grau, 2004). Studies have shown the pushing factors of the rural population in Ethiopia as combination of ineffective and inefficient agricultural marketing system; underdeveloped transport and communications networks; underdeveloped production technologies; limited access of rural households to support services; environmental degradation; lack of participation by rural poor people in decisions that affect their livelihoods (Rural poverty in Ethiopia, 2014). The Ethiopian economy is based mostly on agriculture yielding 46.6% of the total GDP. The other sectors like telecommunications, Banks, Air and Land Transportation services Insurance services are state owned. Despite the high turnover of graduating students from higher educational institutions, the job market availability is too low leaving a huge gap of unemployed educated manpower idle.

As slums expand governments have to put forward policies that improve the living conditions of the slum poor. The continuation of class segregation through social mechanisms, reinforced the coexistence of the rich and poor at much larger concentric circles in Addis Ababa. The lack of laws and regulations pertaining to prostitution coupled with the corrupt police practices added fuel to the fire in maintaining the lack of social order in the city. So if a middle aged man walks with or gives a ride a 14 year old girl fondling her on the road in the middle of the night, so long as the police get their bribes nothing happens to the sugar daddy. Because the rich and poor are juxtaposed in space and time in the city of Addis this has occasionally produced tensions that rose to the level of political riots in the past 20 years. The tensions were partly due to the increased deprivation of the poor, unemployment, the creation of an oligarchy that maintains its survival through corrupt political and social practices. This has led to clashes in recent times in the section of town called paisa targeting the businesses of the oligarch and looting their properties.

One of the hallmarks of slum areas is social exclusion. Social exclusion comprises groups of people that are marginalized from the society due to situations ranging from racism, to unemployment and disabilities (Social Exclusion, 2008). The high level of poverty in slums, the marginalization, and the lack of organization and order in the long run disenfranchises the population residing within these communities, and their participation in the society at large would dwindle. This in turn puts the slum residents as undesirables by other social class members, and leads to rejection and ostracism which can lead to either depression or aggression (Williams, Wesselmann, & Chen, 2007). As these negative social situations continue to play in urban slums, the added high rate of unemployment in these slum communities leads to economic stagnation. Economic stagnation is a situation where the per capita output growth of the economy is at or close to zero for a long period of time, due to situations like high unemployment (Stagnation, 2008). This social exclusion coupled with economic stagnation further compounds the high influx of trafficked women and children to go in to prostitution. In Ethiopia child and women trafficking both domestic and internationally for prostitution has the weakest control by the governments past and present. The trafficking is so out in the open and operates in all the studied cities wearing the cloak of adoption, work agencies in the Middle East and Far East, and/or as scholarships for poor children. There is no follow up mechanism by the government or the embassies in these alleged countries, and in recent times seeing reports of maid suicides on the news in the Arab countries has become a sad norm. These abused women and children if by chance return to their home land they continue the prostitution enterprise in a more complex fashion.

Inequalities Producing the Health Disparities and Increasing Vulnerability

While slums have been part and parcel of city systems in the 19th and 20th century cities, they have escalated in the wake of globalization. Globalization as an economic trend that has integrated world economies and has no boundaries of time, distance, government regulations, cultures; and business systems (Globalization, 2008). The current inequalities in income, education and other social variables have escalated in the wake of globalization and contributed to the formation of slums in many parts of the world. Globalization has been one of the reasons for the increased number of homeless and poor housed populations because of issues of affordability, access to essential services, violation of rights, increasing vulnerability of populations living in slum areas (Kenna, 2008). Now the issue and the pressing question here is asking the following question: since inequality is a multifaceted situation having the faces of race, ethnicity, income, education, gender, class and creed the current globalization, the change in ideology from socialism to the developmental state in Ethiopia, have these contributed to the diversification of prostitution as a coping mechanism of survival? The answer is yes but is more complex than one deliberates, because of the gap in data availability to compare variables. For now we can only provide the observed qualitative analyses on the aspect of prostitution in Ethiopia. As far as international trading is concerned globalization opens up possibilities to the expansion of informal economies especially in slums. Heroin, opium, and cocaine are partly the problem of the poor because their trade is facilitated by the opened international trading channels (Seddon,2008). The quick movement of money through globalized finance system has also resulted in the creation of rogue trading which is characterized by unauthorized type of trading which ends up illegal commodities in slum areas. The slum will thrive in this informal economy where the quality of commodities is not up to the standard escalating the policy and the poverty trap that is rampant in slum areas. The other key cause for the inequalities in slum areas is the uneven labor markets. The current labor markets are not open for unskilled workers. Because these workers don't get good earning jobs they end up in slum areas, and will be engulfed in the slum economies. The high level of poverty seen in slum areas is responsible for the high rate of communicable diseases that are caused by poor water supply, sanitation, housing, and overcrowding. Most slum dwellers are peripheral in government health policies, and thus health services like health facilities, health workers, and medical supplies are not available to slum dwellers. In sub-Saharan Africa overcrowding, substandard housing, poor access to sanitation and clean drinking water are responsible for high prevalence of cholera, dengue fever, malaria, yellow fever, infectious diseases epidemics, child malnutrition, and vaccine preventable childhood illnesses (Ramin, 2009). This opens doors for traffickers who hunt down vulnerable groups due to this underground economy. The vulnerable groups include homeless youth, runaways, and victims of household abuse, illegal immigrants, refugees, and economically marginalized groups. In Ethiopia human trafficking is a legitimate business that is covered by the terms of 'employment agencies', 'adoption agency' and the like and the brokers of this business mobilize vulnerable urban and rural female and child populations living under abject poverty. They promise lucrative employment opportunities for these people and traffic them in prostitution transactions in urban cities in country, and/or traffic them abroad to nearby peninsulas like in the Middle East, Far East and Eastern Europe. Duarte (2012) states the following:

Sex trafficking is far from being an isolated problem. Its causes are intrinsically linked to other social, economic, political, and cultural phenomena, meaning that in several cases, it does not just involve a violation of rights resulting from trafficking. In the specific case of the trafficking of women for the purposes of sexual exploitation, several authors argue that prostitution must be included in these policies, in particular in the legal-normative framework of each country. Barry (1995), one of the founders of the Coalition Against Trafficking in Women and one of the most active voices on this front, argues that sexual exploitation is a political condition and the basis of the subordination of and discrimination against women and the perpetuation of patriarchy. Jeffreys (1999, p. 180) believes that a woman's willingness to engage in prostitution is politically and socially constructed on the basis of poverty, sexual abuse, and women's family obligations. Those who support this position make no distinction between enforced and voluntary prostitution and consider that any concession by the state towards legalization is essentially a concession to constant violations of human rights, dignity, and sexual autonomy (p.258-268).

The Theories and Research Explaining slums

At the core of the issue of prostitution, there is vulnerability, poverty, urbanization and slum formation. The first theory that is linked with slum formation per se is rapid urbanization. Especially in developing countries with the rise in rural-urban migration due to multiple factors like environmental degradation, industrialization, and looking for more opportunities have led to the dramatic increase of poor populations residing in these slum areas. Coupled with the rise in poor capacity of city authorities to cope with the huge influx of migrants in to the cities, the lack of capacity to develop infrastructure in these slum areas, calculate the economic trajectories related to slum expansion such as housing and thus manage the slum areas leads to their expansion(Ooi,Phua, 2007).

The second theory that is attributable to slum formation is population explosion in poor resource countries. This phenomenon increases rural-urban migration increasing the poverty stricken population in slums. On top of the migration issue slums are characterized by high fertility rate and high unmet need for family planning methods, low usage, and poor income (Okech, Wawire, Mburu, 2011).

The third theory linked with slum formation and poor slum management is poor governance. Governments with weak institutions and poor policies have immense challenges in addressing public health issues related to slums. In countries following neo-liberalistic ideologies in the west and Latin America, or those countries that follow the developmental state ideology like china are still struggling with the challenge of slums. Governments with poor mortgage markets for housing loans; or governments with views of increased trivialization of slums; and authoritative government officials that tend to cling to slum demolishing have never solved the slum enigma(Gilbert,2009).

The fourth element related to slums and their public health impact is the lack of adequate address to the environmental and social implications of slums. Highly overcrowded, with the lack of adequate dry and liquid waste disposal, poor clean water supply, poor housing, makes slums the ideal habitats for the proliferation of communicable diseases like diarrhea, and tuberculosis. The high level of exposure to cigarettes and high fat food in these slum areas also makes the population vulnerable to non-communicable catastrophes like lung diseases due to smoking and obesity (Roberts, Patel, Dahab & Mckee, 2013). Social issues like human trafficking, drug trafficking, prostitution, crime and violence are rampant in slum areas (Scalon, 2013).

The fifth theory that is related to slums and negative effects on health is income inequality. Income inequality is characterized by poor investments in education, health, housing, roads, and environmental protection and is directly related with crude mortality (Veenstra, 2002).

V. **Politics**

Ethiopia currently is pursuing the developmental state paradigm. The developmental state is a state led capitalism, where the state leads the macroeconomic planning process. At the same time what is unique about the developmental state is the state apparatus has a complete control on the politics as well as the economics. The developmental state bases its ideology and legitimacy on an ability to promote sustained economic development; its ability to improve the economic conditions of the inhabitants is both the goal of the ruling elite and a means to keep power; it largely depends on private capital and ownership; its bureaucracy is capable of stimulating, shaping, and cooperating with the private sector, identifying industrial projects where the profit-making interests of the private sector coincide with the economic goals of the nation and these interests will normally be common when the private sector invests in projects that increase the technological competence of the nation, often in industries previously dominated by companies based in wealthier countries (Reinart, 2006).

Prostitution is an old profession in Ethiopia. It has been there since time immemorial. However the types of prostitution were not that diversified until recent times. By the end of the Dergue era prostitution has been a well-established profession, and it has been on both domestic and international market. The dominant type of prostitution Ethiopia was the female one. The male type grew significantly to rise to the level of visibility only in the last decade. The socialist regime tried to organize prostitutes in to associations so that they change their way of life. However that was not successful due to financial draw backs on the prostitutes' side. The current regime never addressed the social repercussion of prostitution, and its impact on upcoming young generations overall health and lifestyle. There are challenges to the developmental approaches to reduce the challenge of migration of young Ethiopians to sex work. On one side the high unemployment that is seen in Ethiopia forced the government to open up free migration of Ethiopians in search of 'work' in foreign countries. This would give the government a space to breath and curb political uprisings due to poverty and lack of work. The agencies that facilitate emigration from Ethiopia are labelled as work/employment agencies that connect female Ethiopians for maid work in Middle Eastern countries. On the other hand the majority of the women who went to these foreign countries claim that they never got a decent job in these Middle Eastern countries. Rather they were forced to do sex work and survive. When pressures from international agencies on human trafficking pop up the government shuts down some of these 'employment agencies' by claiming that they are illegal. These employment agencies screen their clients before they send them to Arab countries. Thousands of young women are screened and sent to the embassies of these Arab countries at medical facilities in Addis Ababa. The adoption agencies do the same in exporting children to unidentified individuals and agencies. Poor households are forced to give their kids to unknown people for promises of good life of the child and money. The extreme lack of resources that can be generated from within the country has forced the government to rely on foreign aid. Topping the worldwide list of countries receiving aid from the US, UK, and the World Bank, the nation has been receiving $3.5 billion on average from international donors in recent years, which represents 50 to 60 percent of its national budget (Development Aid to Ethiopia, 2013). Any donor money has strings attached to it, which might not be in the interest of the recipient country's population. True development rarely comes from the outside i.e. top-down development programs administered by governments, international agencies, foundations, or big NGOs rarely work because they're so vulnerable to government corruption, bureaucratic inaction, the distance between the planners and the supposed beneficiaries, and both distrust and a lack of interest on the part of people who live at the grass roots; giveaways breed dependence and self-doubt instead of change, and philanthropy isn't the answer, either; additionally traditional approaches are ill-suited to fight poverty, even the most promising and cost-effective conventional development projects fail to make real headway against poverty because there is never enough money — either from governments or from philanthropists — to take them to scale (Polak, Warwick, 2013). Governments should be innovative enough to address the magnitude of the prevailing social problems like prostitution. Because these social problems are highly intertwined with each other, addressing them requires doable contextualized approach that involves multiple sectors.

The current level of corruption in Ethiopia has set a trend of mafia like transactions the go all the way down to the grass root level of governance. Unless the current Ethiopian government addresses the issues diligently the social maladies that we see today like prostitution, drug use and gaps in implementation of the law will continue to be challenges that will have negative effects both on the society and politics. Economic stride founded on failing social system can lead to the overall collapse of the politics and the state. Prostitution is a form of oppression that incapacitates the individual socially, economically and politically. A political system that has restrictions on the flow of information through free press and media cannot achieve social change, and attain individual liberty. In the current political environment the failure to accept collective failure of the politics to address pressing social issues can result in a political collapse irrespective of the military power of the government. The moth-eaten political ecology where there is so much neglect to human right, morality, and the rise in concentration of poverty mixed with extreme affluence can give way to stark economic segregation, and administrative fragmentation. The current political framework continuously reinforces class and oligarch advantages and social disadvantages. The trend of geographic isolation of the rich from the poor eventually leads to political separation by influencing all branches of the government and its branches where the rich can draw clear lines. This face of corruption that is currently prevailing in both at the government and private level increases the poverty, economic disparity, and further compounds the social issues like prostitution that stem from it. Because of corrupt government practices mortality rates of children or disease prevalence rates like HIV will continue to rise, and the rights of individuals as in women will never get attention. Women rights can only become a reality when class segregations are abated, when there is efficient political decentralization, and when income inequality is addressed effectively. As long as we have these agglomerations of rich that feed on the political advantage of a certain regime amidst the poor, we will never see effective political transference that will benefit the disadvantaged groups. A political system that is skewed towards a certain class, creed, ethnicity, race or religion can only pronounce poverty, inequality, and social exclusion.

The other important aspect of this corrupt political ecology that leads to blurred self-image of people is manifested in the education system. Poor quality education that never strikes that passion of self-worth will end up in producing a confused individual. If a university professor pushes its female student towards transactional sex for 'A's and 'B's, then the country will have a double disadvantage that rises to future cripple of the social and political systems. Education as a wasted global resource will never bear good fruits in societies. It is the most important resource that creates efficient socioeconomic and political systems if it is well cultivated. Working for targeted quota numbers without quality is a waste of time and money. These poorly cultivated graduates are always ill prepared, and that will further compound the class inequalities, low innovativeness, and challenges in politics. The late prime Minister Meles Zenawi has stated in one of his essays (African Development: Dead Ends And New Beginnings , 2011) On the developmental state politics the following:

A related issue has been the need for continuity of policy. Developmental policy is unlikely to transform a poor country into a developed one within the time frame of the typical election cycle. There has to be continuity of policy if there is to be sustained and accelerated economic growth. In a democratic polity uncertainly about the continuity of policy is unavoidable. More damagingly for development, politicians will be unable to think beyond the next election etc. It is argued therefore that the developmental state will have to be undemocratic in order to stay in power long enough to carry out successful development. A critical issue is therefore can such a stable, democratic and at the same time developmental coalition be established in a developing country. It is not difficult based on our analysis so far, to identify who the candidates of such a coalition can be. One group that cannot be part of the coalition is the private sector. One of the defining characteristics of a developmental state is that it must be autonomous from the private sector. It must have the ability and will to reward and punish the private sector actors depending on whether their activities are developmental or rent seeking. It cannot do so if the private sector is in the coalition. Obviously it does not mean that the coalition will have to be hostile to the private sector. It cannot be hostile to the private sector and bring about accelerated development in the context of the market economy. In the end, what the developmental state does will strengthen the value creating part of the private sector more than any other alternative. It only has to be independent from the private sector while at the same time doing things that will punish the rent-seeking part and reward the value creating part of it (p.11-12).

The lifecycle of the developmental state is theorized to have the seed of self-destruct in there because it reaches that level of maturity where there is high level of technological efficiency, high level of manufacturing that leads high demand to large markets which will lead to the demand of free trade (Reinart, 2006). On the other hand looking at the views of Meles the forceful imposition of focusing on economic growth, by shunning democracy aside, leads to further inefficiency of the social system that will be a good ground for the proliferation of social crises like polygonal prostitution. In fact politicians might say prostitution is not a problem at all, it is a type of social service. We should rather focus on other real threats to our government like terrorists etc. Well polygonal prostitution is a huge issue that must be given attention by politicians because it is about human trafficking, it is about human rights (adults, women and children), it has public health ripple effects like in STDs, mental health, substance abuse, homicides, suicides, crimes, it also has ideological repercussions on morals as in deontological versus consequentialist views; and finally cultural and religious repercussions which can branch out into fundamentalist politics.

VI. <u>Legalization, Decriminalization, and Challenges</u>

Prostitution is a multifaceted problem that has serious impacts on society and development in the long term. Human traffickers use coercion and force to recruit young girls with a cloak of well-paying jobs in developed and middle income countries. In Ethiopia prostitution is legal while brothels and pimps are illegal (100 Countries and Their Prostitution Policies, 2014). As far as Ethiopian women are concerned they are trafficked to the Middle East, south East Asia and Eastern Europe through agencies that look for cheap labor in developing poor African countries. In the horn of Africa, the first victims were from south Sudan where civil war was chronic and rampant with lots of displaced women and children in camps and small abandoned towns. The smuggling of women and girls started some 30 years ago through forceful coercion as direct trafficking for prostitution and for child soldiers in the field. After that many agencies in countries like Lebanon, United Arab Emirates, Saudi Arabia and Yemen cropped up smuggling millions of women and children to the Middle East with the cover of workers for housekeeping and catering in these countries. The trafficked women came from a plethora of backgrounds like dysfunctional home, poverty, fleeing conflicts and oppressive governments. In these foreign countries they were abused beyond comprehension, sustaining continuous rape, beatings, and carrying out survival sex. Those who managed to survive they return to their homeland in one piece, but to the most part the majority of these trafficked women either commit suicides, or homicides in self-defense, or sustain long lasting disability through abusive pimps and house owners and return home in coffins or wheel chairs. In Ethiopia the role of the pimp is not that well developed as the western pimp style on the streets. Women and girls practicing street prostitution are trafficked by 'gulbes' which literally means bullies in the street and they are either kidnapped, stacked in a van and taken to abandoned houses to be used by clients and the gulbe will be paid by the client or they pay irregular fees at irregular intervals in the streets to street gangs. (Farley, 2013) states the following:

According to U.N. estimates, approximately 2.5 million people are being trafficked around the world at any given time, 80% of them women and children. Conservative estimates suggest that the sex industry generates some $32 billion annually. However, estimates of income generated from prostitution in one city, Las Vegas, are as high as $5 billion. Today, sex trafficking is a high-tech, globalized, electronic market, and predators are involved at all levels, using the same methods to control prostituted women that batterers use against their victims: minimization and denial of physical violence, economic exploitation, social isolation, verbal abuse, threats and intimidation, physical violence, sexual assault, and captivity. Despite the illogical attempt of some to distinguish prostitution from trafficking, trafficking is simply the global form of prostitution. Sex trafficking may occur within or across international borders, thus women may be either domestically or internationally trafficked or both. Young women are trafficked for sexual use from the countryside to the city, from one part of town to another, and across international borders to wherever there are men who will buy them. Prostitution is an institution akin to slavery, one so intrinsically discriminatory and abusive that it cannot be fixed--only abolished. At the same time, its root causes must be eradicated as well: sex inequality, racism and colonialism, poverty, prostitution tourism, and economic development that destroys traditional ways of living. The conditions that make genuine consent possible are absent from prostitution: physical safety, equal power with johns and pimps, and real alternatives. It is a cruel lie to suggest that decriminalization or legalization will protect anyone in prostitution. Until it is understood that prostitution and trafficking can appear voluntary but are not in reality free choices made from a range of options, it will be difficult to garner adequate support to assist those who wish to escape but have no other choices. Enforcement of international agreements challenging trafficking and prostitution can aid in this effort as can laws challenging men's purchase of sex (p.1).

Beyond human trafficking, prostitution dehumanizes women, children and men who are traded within the circles. It is an exploitative system that commodifies, objectifies and dehumanizes women, men and children reinforcing the subordinate status of the more vulnerable individuals who are more often, women and children; as it serves the instant sexual gratification of the more privileged "clientele" who are mostly male (Enacting the Anti-Prostitution Law, 2009). One of the key driving factors behind the expansion of prostitution is demand. Demand can be expressed in the forms of trading rings, pornographic industry and especially internet pornography has been related as the fire cracker behind the explosion of prostitution globally. Internet has cemented prostitution as a global institution to be reckoned with. Because of that specific demands have risen to the level of 'hunting' young women from specific countries like Ukraine, Ethiopia, Bangladesh, India, the Philippines and the Arabian peninsula for sex slavery and sex trade. One of the key problems even in those countries who have made prostitution illegal is the sloppiness in implementing the law to a point that goes to turning a blind eye to prostitution activities. In Ethiopia prostitution is legal, and that has opened so many platforms to turn it into a thriving business. The prostitution rings in Ethiopia involve policy makers and government officials who utilize the service as regular clients and have become part of the service. This key situation which has been seen in Ethiopia is clearly disclosed by the interviewed sex workers as they tell their clients are the wealthy, upper middle class, and powerful political leaders which use the service either for themselves or for their political guests. Politicizing prostitution for the sake of national income safety net leads to social anarchy and becomes a threat to human rights and health of all society members. This would lead to high level of school drop outs from all school segments, due to sex trafficking and increasing illiteracy and compounding dysfunction in development. As the prostitution industry gains momentum, it becomes so powerful that governments will reach to a point where they have to negotiate with these syndicates in order to maintain political power. At this point government will start to justify prostitution as one of the key means of income for poor segments of the society compounding the anarchy. Countries like the Philippines and Thailand have reached to that stage where the Philippines ranks fourth among countries with the most number of prostituted children and further more a study by the Psychological Trauma Program of the University of the Philippines notes that prostitution may now be the country's fourth largest source of GNP (Stop child porn today, 2014). In Thailand one of the key owners of the country's prostitution brothels and massage houses had run for presidency recently. Results of the study show that socioeconomic disparity, gender inequality, unemployment, illiteracy, culture assassination, 'failing' states and poverty are some of the key factors fostering prostitution. Decriminalizing prostitution has serious ramifications. (Raymond, 2003) states the following:

> Legalization/decriminalization of prostitution is a gift to pimps, traffickers and the sex industry; promotes sex trafficking, does not control the sex industry, it expands it; increases clandestine, hidden, illegal and street prostitution; increases child prostitution; does not protect the women in prostitution; increases the demand for prostitution; it boosts the motivation of men to buy women for sex in a much wider and more permissible range of socially acceptable settings; does not promote women's health; does not enhance women's choice ; and women in systems of prostitution do not want the sex industry legalized or decriminalized (p. 315-332).

Here we have controversial government decisions in countries like the Netherlands where prostitution is one of the key incomes of the country because it is an industry. The Netherlands situation is a key example of moral and ethical dilemma as far as gender rights, and human dignity is concerned. So governments have to choose between their own people and the industry that they are endearing as a source of national income. Some governments are concerned about the rise in unemployed and food insecure populations in their nations and turn a blind eye when high numbers of their fellow women and children are moving out of the country through 'employment agencies' in foreign countries nearby. Prostitution is illegal in the United States. The explanation behind it is because it is part and parcel of organized and ancillary crimes, and it has public health repercussion. Ethiopia has several laws on children; however the implementation is never there. ECPAT (2007) report, states the following on some of the provisions and the gaps:

> Ethiopian law outlines a variety of offences involving sexual acts with children, but falls short of international standards for protecting children from prostitution. As a preliminary matter, Ethiopia has not signed or ratified the *Optional Protocol*. While its few laws related to prostitution address procuring or prostituting a child and provide for enhanced penalties where a crime is committed with the intent to prostitute a child, the *Criminal Code* fails to specifically prohibit the act of having sex with a child for

remuneration. Furthermore, the laws criminalizing sex with children divide children into two categories - those under 13 and those between 13 and 18 years of age - and provide stricter penalties for crimes involving younger children. This approach offers weaker protection to teenage children, who are often the most vulnerable. In several cases, Ethiopian law establishes differences that leave certain categories of children without protection against sexual crimes: for example, its rape law applies only to girls between 13 and 18 years of age, offering no protection to girls under 13 or to boys of any age. On the other hand, positive features of Ethiopian law include stringent penalties for sexual crimes against children; provisions for the prosecution of juridical officers for failing to protect children in certain circumstances from being exploited; and protection so that child victims of certain sexual crimes are not treated as offenders. *Ethiopia acceded to the Convention on the Rights of the Child (CRC) in 1991 but has not signed or ratified its Optional Protocol on the sale of children, child prostitution and child pornography (Optional Protocol). Similarly, Ethiopia has not signed or ratified the Protocol to Prevent, Suppress and Punish Trafficking in Persons, Especially Women and Children (Trafficking Protocol). ILO Convention No. 182 was ratified in 2003. At regional level, Ethiopia ratified but did not sign the African Charter on the Rights and Welfare of the African Child in 2002(p.18-19).*

The most paradoxical situation in Ethiopia is the law on child prostitution because it has not *signed or ratified its Optional Protocol on the sale of children, child prostitution and child pornography*. That continues to make children highly vulnerable for abuse and trafficking. On top of the above issues implementation of the law on child molestation, rape, child trafficking has been a challenge for a long time now. There are no well facilitated institutions with the financial back up to tackle trafficking and abuse. Child prostitution has risen at alarming rate in Ethiopia. Ethiopia has an estimated 6 million orphans and vulnerable children that are at 24/7 risk of being exposed to violence, migration, sexual exploitation, abusive labor, child trafficking, disinheritance, abductions, and unwanted illegal marriages.

Culture and Social exclusion

Prostitution as a result of social exclusion in the context of Ethiopia is an interesting area to explore. Social exclusion is the systematic blockade of individuals and /or communities through a systematic denial of their rights, and opportunities that are otherwise easily and normally available to other groups like housing, civic engagement, employment or due process. When social exclusion involves the individual's social class, standard of living, or educational status it becomes disenfranchisement. It is a subtle form of discrimination that applies in countries like Ethiopia where there is ethnic segregation, socioeconomic disparity, low respect to younger people, lower social position of women, and homophobia. So on the superficial level there are laws and propaganda that would protect these marginalized groups, but beneath the surface due to a strong cultural and religious conservatism and stance there is a continuous tearing of the social fabric which eventually marginalizes individuals from sharing the national cake and normal activities that are prescribed by the Ethiopian society. While prostitution has the dimension of poor choice, in Ethiopia the lack of knowledge about social rights has played deeply through the years. In the regimes that we have witnessed in the past 40 years or so most Ethiopians were systematically and directly usurped of their social rights by repressive governments. (Adamski,2006) states the following:

> Citizens of modern states enjoy a number of rights. Civil and political rights shape individuals' interactions with states' legal and political systems. Economic, social, and cultural rights, on the other hand, address freedoms often exercised in private life. Examples include access to sufficient food, education, health care, and employment. Although economic, social, and cultural rights offer different guarantees than do civil and political rights, the international community treats them as indivisible. Because they reinforce each other, together they help to ensure social justice. For example, without the political right of free association, the economic right to form unions would be meaningless. Equally, the social and cultural right to an education would be worthless to those imprisoned because they do not enjoy the civil right to be free from arbitrary detention. The Universal Declaration of Human Rights passed in the UN General Assembly without a dissenting vote. It outlined thirty principles basic to human development and dignity, including not only economic, social, and cultural rights, but also civil and political rights. Following passage, the body asked the Commission on Human Rights to put the general language of the Universal Declaration into a form legally binding on states with a treaty outlining specific rights and

their implementation. This eventually led to two treaties: the International Covenant on Cultural and Political Rights (ICCPR) and the International Covenant on Economic, Social, and Cultural Rights (ICESCR). Signaling the political disputes involved, neither covenant passed the General Assembly until 1966. Each required an additional ten years, until 1976, before receiving the thirty-five votes required for ratification (p.14-21).
Social rights is interrelated and with economic and cultural rights. Material deprivation and poverty that is rampant in the Ethiopian society seeped through time into the already disarrayed type of social decisions that led to negative cultural stances towards children, women, and young people. Whooping a child until they get bruises on their bodies is considered normal disciplining, while it seriously affected the child's psyche and the family relationships. Misquoting holy books towards the benefits of the oppressor have also played significantly in the synthesis of social exclusion in the Ethiopian society. Although it's always argued that there is a significant social engagement in the Ethiopian society, what we see and observe at the present time point towards the negative side of it. For that matter there have been no quantifiable measurements in the past on the issue of social engagement and participation. Usurping social and cultural rights would result in weak or negative participation in social activities, and that would result in loopholes of social sanctioning of attitudes and practices that would result in criminal activities like prostitution. In normal circumstances social engagement reinforces social capital. (Glanville,2009) states the following:
Social capital refers to social relationships that have the capacity to enhance the achievement of one's goals. Social relationships are viewed as investments, whether the investment is conscious or unconscious. The fundamental insight of the idea of social capital is that life chances are influenced by the social resources available through social networks. This fundamental insight, coupled with the generality of the definition, has contributed to social capital's widespread appeal across a variety of research areas, including research on child and adolescent development. Social capital is believed to aid in the achievement of both individual goals, such as educational attainment and achievement, and collective goals, such as safe neighborhoods that promote healthy psychosocial development (p. 442-446).
Looking at the premises of social capital that are necessary to expand a healthy society with good control crimes, the urban context of Ethiopia at least in the

studied cities shows a negative refractory image of social capital that is unhelpful towards progress. In fact it shows a gray area in the normative integration and social cohesion between individuals and their community in the studied cities. This shows the once upon a time strong social cohesion in certain areas of sociological context like family values, or the urban common sense that we used to see have faded out and have been replaced by new rules and new games. This context clearly shows that entities like polygonal prostitution result from extensive social exclusion that is founded on economic changes, technology (internet, Facebook, porn), social policy, individual factors like gender, ethnicity, race, and age; and the mafia nature of businesses.

In Ethiopian urban areas as far as social exclusion is concerned in polygonal prostitution individual exclusion plays a significant role. Individual exclusion is marked by the lack of meaningful participation in the society. This especially holds true on Ethiopian women. The lack of cultivating Ethiopian women with good quality education has resulted in what Paulo Friere called 'the pedagogy of the oppressed' and the culture of silence. This has further facilitated their individual exclusion in a marked way to a point educated women or those having a middle class life considered and practiced prostitution as a good alternative income generating activity. (Freire, 2008) states the following:

> Freire emphasized the dialectic relationship between theory and practice, which is expressed through three generative themes in his work: *concientization, dialogic learning*, and his critique of the *banking approach* to education. Underpinning these three generative themes is a student-centered system of learning that challenges how knowledge is constructed in the formal education system and in society at large. Freire's student-centered approach stands in stark contrast to conventional educational practice, which he referred to as the "banking approach" to education. He argued that conventional learning was the tool of the elite because it treated students as objects upon which knowledge is "deposited." Genuine learning, for Freire, could only be achieved through lived experience, critical reflection, and praxis (Aronowitz 1993, p. 9).The idea that "experiences are lived and not transplanted" is a central tenet of Freire's philosophy (Gadotti 1994, p. 46). Concientization is the key process by which students develop a critical awareness of the world based on the concrete experience of their everyday lives. The development of critical awareness through concientization alters

power relations between students and teachers, the colonized and the colonizer, thereby transforming objects of knowledge into historical subjects (Freire 1997). Freire proposed that a dialogical theory of action based on communication and cooperation was necessary not only for understanding the mediating role of historical, colonial, and class relations (concientization), but also for the active work of changing them. Dialogic action challenges mediating social realities by posing them as problems that can be analyzed critically by those who have direct experience of them. Dialogue becomes a form of collective praxis directly concerned with unveiling inequitable conditions obscured by the ruling classes (p.201-202).

The male dominated Ethiopian culture in almost all regions has been a threat to the rights of women and children at all times despite some of the strides in asserting women and children rights in the last three decades. The rights of Ethiopian women has a long way to go to the level of equality. When we compare women rights in Addis Ababa with the rest of the regions, women rights in Addis seem to be better than the regions in many aspects like marriage, abduction, education, and legal protection. In the rural areas of Ethiopia abduction, forced and arranged marriage, widow inheritance, female genital mutilation are rampant; access to health care, the level of opportunities in education, economics, social and other political participations is at the stage of infancy. Because of the deep cultural values that are rooted in religious beliefs of orthodox Christianity and Muslim religions women are still struggling as far as getting equal opportunities is concerned. This deprivation ultimately has been the crux of the displacement of women from their households and migrate to the city and abroad in search of good life and jobs. It is because of this very situation predators of trafficking take advantage of these marginalized women and children.

Here are some facts from Ethiopia DHS (2011) on the socioeconomic characteristics of urban women:

28.3% have no education, 9.4% have completed secondary education; 27 percent of children age 5-14 in Ethiopia are involved in child labor; 6% of urban women access all print and mass media once a week; 49.9% urban women are employed; and 35.8% of households are headed by females.

The rural areas have more marked female disadvantage when compared to socioeconomic indicators. In most rural communities of Ethiopia communities are organized into smaller villages, there is high social value and network, in which all traditional rewards and sanctions are mediated through. Because of the high social control, and oppressive moral and ethical codes on women and children in the rural areas of Ethiopia most women and children run away for fear of their lives and in search of positive livelihood. However on the other hand when they reach the urban areas they are hit by cultural shock, and they are drawn into the negative depths of urbanism. Urbanism has organization problems socially and is different from rural community structures in the sense that there is high level of individualism, has its own language, shifts rural people personality in a negative way, alienation, and deviance. No one has worked more on urbanism as Louis Wirth. Wirth, Louis 1897-1952. (2003) states the following:

> In "Urbanism as a Way of Life," Wirth defines cities as large areas populated densely by heterogeneous people. He views urbanism, the growth of cities at the expense of rural areas, to be a continuing trend in modern society. Moreover, as his theory of "new urbanism" states, the process of living in a large city changes the way people relate to one another. Primary relationships are not as important as secondary relationships, and secondary relationships tend to be superficial and revolve around the limited roles people assume: banker, cashier, doorman, mail carrier, etc. Wirth explains how and why people tend to group themselves together—by religion or race—based on the laws of homogeneity, a fact that leads to segmentation of urban life, which although it may be detrimental to community involvement, often leads to a satisfactory measure of personal freedom (p. 439-441).

The alienation and the negative impersonation would further take vulnerable groups like women and children who are engaged in sex work and polygonal prostitution in to a whirl pool of poverty and abuse that takes them under. That's why we need to redefine vulnerability, most at risk groups, and the high value of organizing marginalized groups in to communities in urban areas so that they can change their life styles.

VII. Addressing Vulnerability and Recommendations

Polygonal prostitution calls for an interdisciplinary approach to address it to the context of location i.e. whether it's urban or rural, culture, level of infrastructure development, economics and politics. Up to know public health interventions have associated prostitution only to certain types of sex workers who are labeled as most at risk populations in HIV interventions. However we have seen from this study that prostitution is highly polygonal, multi-layered and literally in countries like Ethiopia there is no one exempt of sex work practice for money, which is highly alarming. This study observed the context of urban areas in Ethiopia and the frameworks that are going to be addressed here in should be contextualized to urban settings. Because prostitution get more complex and diversified as urban population grows, and the economy booms, it will be very difficult to isolate certain segment of sex workers and or populations to be the focus of public health interventions. That might help in the short term like decreasing HIV prevalence, but it will not solve the bigger issue in the long term. In such situations public health interventions should be part and parcel of interdisciplinary teams made up of researchers and practitioners from a wide range of academic fields like law, psychology, anthropology, urban planning, medicine, public health, public policy, sociology, economics, education, criminal justice, women's health, environmental justice, and human rights. The magic in this interdisciplinary framework is to synthesis new frame works and research theories for a change of paradigm. The interdisciplinary approach will have the advantages of integrating and reconciling research and practice that aims to improve population health and that seeks to promote social justice; researchers can contribute to the reintegration of these two concerns by studying the linkages among allocation of power, political and social processes, and health outcomes; and it will abolish that critical prevailing divorce of public health research from its roots in social justice which has been focused more on documenting individual-level risk factors and studying the impact of various techniques for financing and delivering medical care (Freudenberg, Klitzman, Saegert, 2000).

In order to address polygonal prostitution and to help identification of vulnerable groups the factors playing in diversifying polygonal prostitution should be addressed. The following socio economic, environmental and political determiners are related to polygonal prostitution:

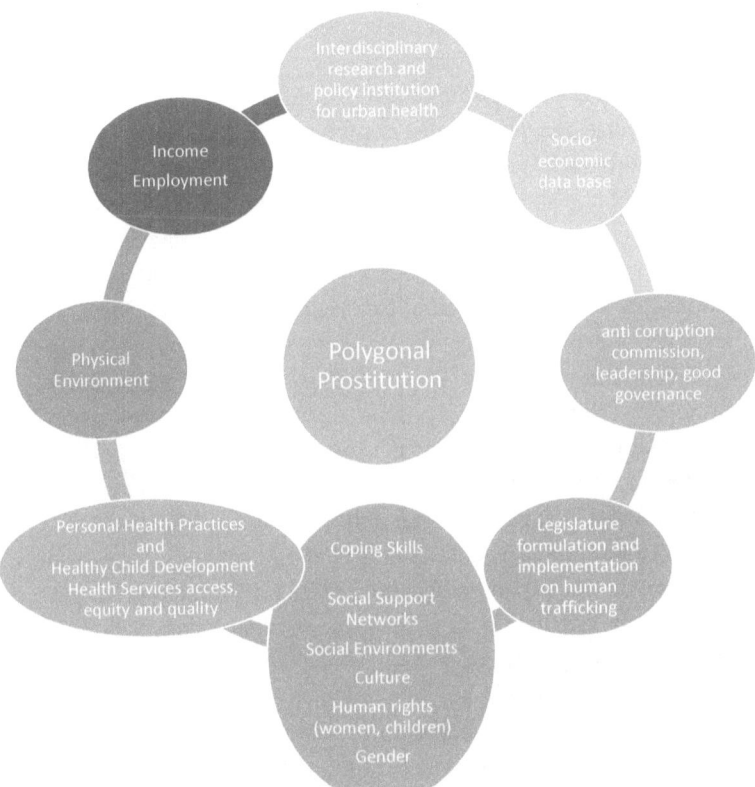

Figure 12. Factors Related to Polygonal Prostitution
The policies in solving the slum versus public health paradigm should revolve around in solving income inequalities, and expanding socioeconomic development. Income inequalities produce disintegration, which in turn produces unhealthy societies, with low life expectancy and they are the results of low investments in human, physical and cultural capital (Beckfield, 2004). It is highly imperative that government policies tackle the central issues of income inequalities that lead to the expansion of slums, crime, violence, prostitution, human trafficking, substance abuse and the rise of morbidities and mortalities related to mental health. Parallel to these issues policies should emphasize in maximizing urban infrastructure and housing development, addressing issues of population explosion, and pushing factors for the intense rural-urban migration. Cities should approach the issue of

slum expansion through an integrated strategic inter-sectoral approach, through joint planning and execution of policy agenda.

Cities are characterized by complex biological, social, and physical systems in which organisms; diverse and dense populations; population heterogeneity ; high levels of inequality; programs segregated by sectors in municipalities; they have a rich array of social and human resources; and are characterized by high level of dynamism (Freudenberg, Klitzman, Saegert, 2000). Because of this fact no one discipline is able to address the complex socioeconomic, political, and environmental issues that are linked with urban health. This calls for a broader interdisciplinary research and planning team to always plan and monitor urban health in a continuum through effective joint planning and implementation of urban health programs. Side by side with infrastructure development like housing, sewage system building, road and rail way construction, the development of community centers, health facility expansion alongside of capacity building in human resources, technology, and financing in health care is imperative.

Slum areas grow due to different socioeconomic and political reasons. Population explosion, increased rural-urban migration, poor economic development, and economic vicious cycle, conflicts, poor governance and leadership, widening wealth gap and income inequality can be some of the key reasons for slum formation. Many countries have tried to improve the living conditions in slum areas by upgrading, removing or relocating them. However, since slum areas are stark reflections of how the economy is performing finding solutions for income inequality by expanding the middle class, increasing wages, investing in higher education, urban infrastructure development, increasing production of commodities that are available for the middle class, improving government investments, improving housing and expanding economies and strengthening a virtuous cycle in the economy. Slum improvement is a multifaceted issue that involves all sectors and stakeholders in infrastructure development like transport, education, communication, energy, water supply, sanitation and waste disposal, health sector development, public security like policing and crime prevention and control, improving the economy, and most of all efficient political leadership and governance.

To alleviate the infrastructure and health problems in slum areas many efforts have been undertaken by different governments around the world. Some of the efforts included slum upgrading, slum removal, slum relocation and infrastructure development. All of these efforts have produced different results in different locations around the world. Because slums have issues of land tenuring, and legitimacy of property especially land most upgrading efforts have produced negative unsustainable results (Werlin,1999). Slum relocation and removal has also similar negative results that are associated land tenure, and funding for housing construction (Administration of Urban Planning and Community and Rural Development, 2011). One promising area to alleviate and address the problem of urban slums is urban infrastructure development and housing. By diversifying funding sources for infrastructure development by involving stakeholders like the private sector with governments positive results can be achieved (Kyvelou, Karaiskou, 2006).

To boost income and employment for vulnerable populations emphasis on education should be given depending on the interest areas. This would be easily achieved through income redistribution schemes where progressive tax schemes will be employed and the rich will be taxed higher than the low income earners. The revenue can be used to educate those needy vulnerable populations so as they build skills to cope within the economic system. Education can be a part of skill building and opening up investment opportunities for affected populations. Income redistribution schemes are part of strengthening public service systems. By redistributing income across the board the economic stratification decreases, and there will be lesser social inequality and a greater social justice. I also further emphasize what Amartya Sen has theorized on the goods of distributive justice further. (Justice, Distributive, 2008) states the following on this issue:

More recently, economist and philosopher Amartya Sen has sought to join concerns for the well-being of individuals with the disadvantages they acquire as members of a discriminated class. Every individual, Sen argues, is an assemblage of specific capabilities. To provide equality of opportunity, interventions need to provide resources that strengthen an individual's capabilities to function in a manner she finds conducive to her well-being. Her freedom of choice is preserved, along with her desire to live her particular life. To get resources to individuals with discernible capability deficits, Sen argues strongly that resources must be targeted to specific kinds of persons rather than simply added to a society's macroeconomic mix. Hence, the individuality of deficits is matched with ameliorative social programs based upon the realization that deficits can be identical, person to person. Gender inequality, for instance, commonly affects certain women who could be grouped into a class, while this same class could differ from other women with another provenance (p. 239-241).

Sen's argument is another possible avenue in specifically identifying more vulnerable groups' skill sets and inclinations and support them change their lives. I completely oppose what Herrnstein and Murray have stated in this aspect. Their argument is entirely based on intelligence quotient (IQ), which states that differences in intelligence lead to the formation of elite based upon their superior abilities, and subaltern status for those less fortunate(Justice, Distributive, 2008). The IQ has its own issues of bias like age, and contextualization. In fact the Murray-Herrnstein theory is nothing but another platform to strengthen the social construct of racism that was prematurely designed by Carl Linnaeus through scientifically tainted mechanisms.

Establishment of social data base helps national and regional registry of those working in polygonal prostitution. Open forums on the issue of polygonal prostitution helps to address the magnitude of the issue, its societal damages, and mechanisms to curb it. When carrying out this study it was very difficult to trace some of the types of prostitution just because there were few studies carried out in this aspect. Censuses should critically consider putting out preliminary data surrounding social maladies like prostitution, substance use, crime and violence data as part of the national and regional social data base.

Corruption should be thoroughly addressed in developing countries. Ethiopia is hit by corruption epidemic that rose to epic proportions by government officials and members of the private sector. The key side effects of corruption are distortion of labor markets, misallocation of resources, alteration of income distribution, distortion of public-sector projects, and declining investments. The more bribe is given to officials, the higher will be the next price of bribe, and it creates lack of confidence in entrepreneurs, and creates mal-allocation of resources towards unnecessary projects. It creates and facilitates a vicious cycle of poor governance, escalation of poverty, increases disadvantages of the poor, and creates poor delivery of services like education and health care. To solve issues of corruption radical measures and following them through thorough monitoring is imperative. (Corruption, 2008) states the following on these critical solutions:

Democratic institutions that allow a fair and wide participation of the public in the selection of public officials create a barrier against the spread of corruption. Offending officials will be identified and punished by the judicial system or at least not reappointed to their positions. A free press accompanied by an independent and honest judiciary forms another barrier that prevents corruption. Institutions and a variety of organizations that represent specific interest groups in a given society will exert influence on public officials if corruption is likely to harm that society. In so far as politicians derive their authority and appointment from their constituencies, these institutions form another line of defense against corruption. Once in power, however, public officials will have means to render these control mechanisms less effective. A dictator usually succeeds in rendering them completely ineffective; hence, dictatorial regimes are often more corrupt than democratic ones. Democratic leaders are restricted by the strengths of the institutions in a society that constrains behavior of public officials (p. 143-146). Most African countries have anti-corruption laws on paper. The implementation is however literally non-existent. Because of this wide spread political gray area social maladies are not solved appropriately through the right institutions along the continuum. Instead only those problems that only have repercussions to the political power are addressed in a fire extinguishing manner.

The issue of women is a work in progress in Ethiopia. In the past two decades positive strides in certain areas of gender issues like abduction, abortion laws, gender based violence and improving equality have been observed. However, when we see the overall situation women in politics, economic advancement, education, and trafficking in Ethiopia it is at the stage of infancy.

Here are some hard facts about global trends on women (The World Women, 2010):

Two thirds of the 774 million adult illiterates worldwide are women – the same proportion for

the past 20 years and across most regions. In tertiary enrolment, men's dominance has been reversed globally and gender disparities favor women, except in sub-Saharan Africa and Southern and Western Asia. Women are predominantly and increasingly employed in the services sector. Vulnerable employment – own-account work and contributing family work – is prevalent in many countries in Africa and Asia, especially among women. The informal sector is an important source of employment for both women and men in the less developed regions but more so for women. Occupational segregation and gender wage gaps continue to persist in all regions. Part-time employment is common for women in most of the more developed regions and some less developed regions, and it is increasing almost everywhere for both women and men. Women spend at least twice as much time as men on domestic work, and when all work – paid and unpaid – is considered, women work longer hours than men do. Becoming the Head of State or Head of Government remains elusive for women, with only 14 women in the world currently holding either position. In just 23 countries do women comprise a critical mass – over 30 per cent – in the lower or single house of their national parliament. Worldwide on average only one in six cabinet ministers is a woman. Women are highly underrepresented in decision-making positions at local government levels. In the private sector, women continue to be severely underrepresented in the top decision making positions. Only 13 of the 500 largest corporations in the world have a female Chief Executive Officer. Violence against women is a universal phenomenon. Women are subjected to different forms of violence – physical, sexual, psychological and economic – both within and outside their homes. Rates of women experiencing physical violence at least once in their lifetime vary from several per cent to over 59 per cent depending on where they live. Current statistical measurements of violence against women provide a limited source of information, and statistical definitions and classifications require more work and harmonization at the international level. Female genital mutilation – the most harmful mass perpetuation of violence against women – shows a slight decline. In many regions of the world longstanding customs put considerable pressure on women to accept abuse. Households of lone mothers with young children are more likely to be poor than households of lone fathers with young children. Women are more likely to be poor than men when living in one-person households in many countries from both the more developed and the less developed regions. Women are overrepresented among the older poor in the more developed regions. Existing statutory and customary laws limit women's access to land and other types of property in most countries in Africa and about half the countries in Asia. Fewer women than men have cash income in the less developed regions, and a significant proportion of married women have no say in how their cash earnings are spent. Married women from the less developed regions do not fully participate in intra-household decision-making on spending, particularly in African countries and in poorer households (p. 19-157).

Creating job opportunities by giving emphasis to the expansion of the middle class, through the expansion of tertiary education is imperative in order to decrease the trafficking, improve self-sustenance, liberty, and equality. Putting out legislature and implementing them effectively especially on individuals and institutions that progress women and child trafficking both domestically and internationally is fundamental in ensuring human rights. Elimination of cultural practices that lead to abuse of women and children through intensive community and public dialogue is highly essential so as to eradicate abuse, victimization during violence, and minimize crimes.

Health wise some of the complications of prostitution in their health like unwanted pregnancy, abortion, unwanted children, HIV/AIDS, Drug addiction, ovarian issues, break the parts of the body and some other psychological and social issues and in respect of customers' behavior like refusing of payment, kidnapping and sexual assault, beating, forcefully abusive acts, violence, drug abusing, threatening, tie and sexual abuse (Qayyum et.al, 2013). As a public health worker, targeting the unique and the vulnerable most at risk groups (MARPs) for public health interventions like HIV/AIDS programs has its own formula when targeting specific vulnerable groups . Most non-governmental and governmental reproductive health and HIV prevention and control programs when they design reproductive health and HIV interventions that relate to most at risk groups should include the issues of polygonal prostitution. Public health programs if they address polygonal prostitution they will maximize the coverage of health care services to otherwise needy and vulnerable populations.

References

Adamski, J. (2006). Economic, Social, and Cultural Rights. In C. N. Tate (Ed.), *Governments of the World: A Global Guide to Citizens' Rights and Responsibilities* (Vol. 2, pp. 14-21). Detroit: Macmillan Reference USA. Retrieved from http://go.galegroup.com/ps/i.do?id=GALE%7CCX3447400104&v=2.1&u=lirn410 86&it=r&p=GVRL&sw=w&asid=77e035555e522aa4be942f85fbb24a8c

Andreescu, C., & Zaharie, C. G. (2014). DOES PROSTITUTION HAVE ECONOMIC CAUSES? *Romanian Economic and Business Review, 9*(1), 107-112. Retrieved from http://search.proquest.com/docview/1537944919?accountid=160851

Aral, S. O., Hughes, J. P., Stoner, B., Whittington, W., & al, e. (1999). Sexual mixing patterns in the spread of gonococcal and chlamydial infections. *American Journal of Public Health, 89*(6), 825-33. Retrieved from http://search.proquest.com/docview/215090324?accountid=160851

aspe.hhs.gov. (2013). 2013 Poverty guidelines. Retrieved from http://aspe.hhs.gov/poverty/13poverty.cfm

Beckfield, J. (2004). Does income inequality harm health? new cross-national evidence*. *Journal of Health and Social Behavior, 45*(3), 231-48. Retrieved from http://search.proquest.com/docview/201663965?accountid=160851

Carballo-diéguez, A., Balan, I., Dolezal, C., & Mello, M. B. (2012). Recalled sexual experiences in childhood with older partners: A study of brazilian men who have sex with men and male-to-female transgender persons. *Archives of Sexual Behavior, 41*(2), 363-76. doi:http://dx.doi.org/10.1007/s10508-011-9748-y

Church, S., Henderson, M., Barnard, M., & Hart, G. (2001). Violence by clients towards female prostitutes in different work settings: Questionnaire survey. *British Medical Journal, 322*(7285), 524-5. Retrieved from http://search.proquest.com/docview/204013666?accountid=160851

Corruption. (2008). In W. A. Darity, Jr. (Ed.), *International Encyclopedia of the Social Sciences* (2nd ed., Vol. 2, pp. 143-146). Detroit: Macmillan Reference USA. Retrieved from http://go.galegroup.com/ps/i.do?id=GALE%7CCX3045300467&v=2.1&u=lirn41086&it=r&p=GVRL&sw=w&asid=6956bb3bff5a6572cd7499ed69ecce50

Debroy, B. (2006). Human Development Indicators. In S. Wolpert (Ed.), *Encyclopedia of India* (Vol. 2, pp. 215-219). Detroit: Charles Scribner's Sons. Retrieved from http://go.galegroup.com/ps/i.do?id=GALE%7CCX3446500274&v=2.1&u=lirn41086&it=r&p=GVRL&sw=w&asid=106f2358a4eea4b96a2b455cdb26c2cd

Duarte, M. (2012). Prostitution and trafficking in portugal: Legislation, policy, and claims. *Sexuality Research & Social Policy, 9*(3), 258-268. doi:http://dx.doi.org/10.1007/s13178-012-0093-2

Eyo, E. O., & Ogo, U. I. (2013). ENVIRONMENTAL IMPLICATION OF OVER POPULATION AND RURAL-URBAN MIGRATION ON DEVELOPMENT IN NIGERIA. *Academic Research International, 4*(6), 261-271. Retrieved from http://search.proquest.com/docview/1515639447?accountid=160851

Farley, M. (2013). Human Trafficking and Prostitution. Retrieved from http://www.psysr.org/issues/trafficking/farley.php

Feminism. (2008). In W. A. Darity, Jr. (Ed.), *International Encyclopedia of the Social Sciences* (2nd ed., Vol. 3, pp. 119-122). Detroit: Macmillan Reference USA. Retrieved from http://go.galegroup.com/ps/i.do?id=GALE%7CCX3045300810&v=2.1&u=lirn410 86&it=r&p=GVRL&sw=w&asid=0ed21fc1439ffeaa216612db32bc24ef

Flowers, R. Barri (2010). *Street kids: the lives of runaway and thrown away teens*. McFarland. pp. 110–112. ISBN 0-7864-4137-2.

Freire, Paulo. (2008). In W. A. Darity, Jr. (Ed.), *International Encyclopedia of the Social Sciences* (2nd ed., Vol. 3, pp. 201-202). Detroit: Macmillan Reference USA. Retrieved from http://go.galegroup.com/ps/i.do?id=GALE%7CCX3045300862&v=2.1&u=lirn410 86&it=r&p=GVRL&sw=w&asid=aa84d05d88e4e22131fda42e473ec0bb

FREUDENBERG, N., KLITZMAN, S., & SAEGERT, S. (2009). *URBAN HEALTH AND SOCIETY; Interdisciplinary Approaches to Research and Practice* (Rev ed.). San Francisco, CA: John Wiley & Sons.

Gilbert, A. (2009). Extreme thinking about slums and slum dwellers: A critique. *The SAIS Review of International Affairs, 29*(1), 35-48. Retrieved from http://search.proquest.com/docview/231351792?accountid=160851

Glanville, J. L. (2009). Social Capital. In D. Carr (Ed.), *Encyclopedia of the Life Course and Human Development* (Vol. 1, pp. 442-446). Detroit: Macmillan Reference USA. Retrieved from http://go.galegroup.com/ps/i.do?id=GALE%7CCX3273000117&v=2.1&u=lirn410 86&it=r&p=GVRL&sw=w&asid=8d60ca4d8ef7f45b4355f055f7b268bb

Globalization. (2008). In *Everyday Finance* (Vol. 1, pp. 225-228). Detroit: Gale. Retrieved from http://go.galegroup.com/ps/i.do?id=GALE%7CCX2830600094&v=2.1&u=lirn410 86&it=r&p=GVRL&sw=w&asid=47f6133554c2b3d1a547416ddf86bad5

Gregson, S., Nyamukapa, C. A., Garnett, G. P., Mason, P. R., & al, e. (2002). Sexual mixing patterns and sex-differentials in teenage exposure to HIV infection in rural Zimbabwe. *The Lancet, 359*(9321), 1896-903. Retrieved from http://search.proquest.com/docview/199013676?accountid=160851

hdr.undp.org. (2013). Human Development Report 2013. The Rise of the South:Human Progress in a Diverse World. Retrieved from http://hdr.undp.org/sites/default/files/Country-Profiles/ETH.pdf

Heidi Hoefinger. *Negotiating Intimacy: Transactional Sex and Relationships Among Cambodian Professional Girlfriends*, PhD dissertation, Goldsmiths, University of London, July 2010

http://eprints.gold.ac.uk/3419/1/Heidi_Hoefinger_Transactional_Sex_PhD_2010.pdf

Hossain, M., M.Sc, Zimmerman, C., PhD., Abas, Melanie,M.D., M.Sc, Light, M., M.Sc, & Watts, C., PhD. (2010). The relationship of trauma to mental disorders among trafficked and sexually exploited girls and women. *American Journal of Public Health, 100*(12), 2442-9. Retrieved from http://search.proquest.com/docview/804340292?accountid=160851

Hovey, J. (2007). Homophobia. In F. Malti-Douglas (Ed.), *Encyclopedia of Sex and Gender* (Vol. 2, pp. 715-717). Detroit: Macmillan Reference USA. Retrieved from

http://go.galegroup.com/ps/i.do?id=GALE%7CCX2896200303&v=2.1&u=lirn41086&it=r&p=GVRL&sw=w&asid=3d98f08b481a6bb3ccab8e8d36451de7

James, W. H. (2004). The Cause(s) of the Fraternal Birth Order Effect in Male Homosexuality. *Journal of Biosocial Science, 36*(1), 51-9, 61-2. Retrieved from http://search.proquest.com/docview/203943984?accountid=160851

Justice, Distributive. (2008). In W. A. Darity, Jr. (Ed.), *International Encyclopedia of the Social Sciences* (2nd ed., Vol. 4, pp. 239-241). Detroit: Macmillan Reference USA. Retrieved from

http://go.galegroup.com/ps/i.do?id=GALE%7CCX3045301236&v=2.1&u=lirn41086&it=r&p=GVRL&sw=w&asid=5e8d8e69e6c712087c3982634a3dc37a

Kempe,Ronald Hope,,Sr. (1999). Managing rapid urbanization in africa: Some aspects of policy. *Journal of Third World Studies, 16*(2), 47-59. Retrieved from http://search.proquest.com/docview/233191025?accountid=160851

Kenna, P. (2008). Globalization and housing rights. *Indiana Journal of Global Legal Studies, 15*(2), 397-469. Retrieved from http://search.proquest.com/docview/236687673?accountid=160851

Khatun, F., Rasheed, S., Moran, A. C., Alam, A. M., Shomik, M. S., Sultana, M., . . . Bhuiya, A. (2012). Causes of neonatal and maternal deaths in dhaka slums:

Implications for service delivery. *BMC Public Health, 12*, 84. doi:http://dx.doi.org/10.1186/1471-2458-12-84

Kyvelou, S., & Karaiskou, E. (2006). Urban development through PPPs in the euro-mediterranean region. *Management of Environmental Quality, 17*(5), 599-610. doi:http://dx.doi.org/10.1108/14777830610684567

Mark Hunter. "The Materiality of Everyday Sex: Thinking Beyond 'Prostitution'", African Studies, 61(1): 99-120.

newswire.uark.edu. (2014). Affluent, Educated Women May Choose Sexual Prostitution. Retrieved from http://newswire.uark.edu/articles/16181/affluent-educated-women-may-choose-sexual-prostitution

Okech, T. C., Wawire, N. W., & Mburu, T. K. (2011). Contraceptive use among women of reproductive age in kenya's city slums. *International Journal of Business and Social Science, 2*(1) Retrieved from http://search.proquest.com/docview/904526245?accountid=160851

Ompad, D. C., Strathdee, S. A., Celentano, D. D., Latkin, C., Poduska, J. M., Kellam, S. G., & Ialongo, N. S. (2006). Predictors of early initiation of vaginal and oral sex among urban young adults in baltimore, maryland. *Archives of Sexual Behavior, 35*(1), 53-65. doi:http://dx.doi.org/10.1007/s10508-006-8994-x

Ooi, G. L., & Phua, K. H. (2007). Urbanization and slum formation. *Journal of Urban Health, 84*, 27-34. doi:http://dx.doi.org/10.1007/s11524-007-9167-5

pcw.gov.ph. (2009). Enacting the Anti-Prostitution Law (Amending Articles 202 and 341 of the Revised Penal Code). Retrieved from http://pcw.gov.ph/wpla/anti-prostitution

Polak, Paul & Warwick, Mal. (© 2013). The business solution to poverty: designing products and services for three billion new customers. [Books24x7 version] Available from http://common.books24x7.com.ezp-02.lirn.net/toc.aspx?bookid=56083.

Potterat, J. J., Rothenburg, R. B., Muth, S. Q., Darrow, W. W., & Phillips-Plummer, L. (1998). Pathways to prostitution: The chronology of sexual and drug abuse milestones. *The Journal of Sex Research, 35*(4), 333-340. Retrieved from http://search.proquest.com/docview/215280756?accountid=160851

prostitution.procon.org. (2009). 100 Countries and Their Prostitution Policies. Retrieved from http://prostitution.procon.org/view.resource.php?resourceID=000772#ethiopia6.http://prostitution.procon.org/view.resource.php?resourceID=000772#ethiopia

Qayyum, S., Iqbal, M. M. A., Akhtar, A., Hayat, A., Janjua, I. M., & Tabassum, S. (2013). CAUSES AND DECISION OF WOMEN'S INVOLVEMENT INTO PROSTITUTION AND ITS CONSEQUENCES IN PUNJAB, PAKISTAN. *Academic Research International, 4*(5), 398-411. Retrieved from http://search.proquest.com/docview/1447235824?accountid=160851

Ramin, B. (2009). Slums, climate change and human health in sub-saharan africa. *World Health Organization.Bulletin of the World Health Organization, 87*(12), 886. Retrieved from http://search.proquest.com/docview/229662226?accountid=160851

Raymond, J. G. (2003). Ten Reasons for Not Legalizing Prostitution. Retrieved from http://www.embracedignity.org/uploads/10Reasons.pdf

Reinart, E. (2006). Developmental State, Concept of the. In J. J. McCusker (Ed.), *History of World Trade Since 1450* (Vol. 1, pp. 191-193). Detroit: Macmillan Reference USA. Retrieved from http://go.galegroup.com/ps/i.do?id=GALE%7CCX3447600113&v=2.1&u=lirn410 86&it=r&p=GVRL&sw=w&asid=ccb5f31607806d2abcaef09f175c98f5

Resl, V., Kumpova, M., Cerna, L., Novak, M., & Pazdiora, P. (2003). Prevalence of STDs among prostitutes in czech border areas with germany in 1997-2001 assessed in project "jana". *Sexually Transmitted Infections, 79*(6), 452-E3. Retrieved from http://search.proquest.com/docview/194381069?accountid=160851

Roberts, B., Patel, P., Dahab, M., & Mckee, M. (2013). The arab spring: Confronting the challenge of non-communicable disease. *Journal of Public Health Policy, 34*(2), 345-52. doi:http://dx.doi.org/10.1057/jphp.2013.14

Sadler, Philip. (© 2010). Sustainable growth in a post-scarcity world: consumption, demand, and the poverty penalty. [Books24x7 version] Available from http://common.books24x7.com.ezp-01.lirn.net/toc.aspx?bookid=37504.

Scalon, C. (2013). Rio and its slums: The future at the crossroads. *Contemporary Sociology, 42*(2), 199-202. Retrieved from http://search.proquest.com/docview/1465307552?accountid=160851

Schrader, S. (2007). Planet of slums. *NACLA Report on the Americas, 40*(1), 46-49. Retrieved from http://search.proquest.com/docview/202677740?accountid=160851

Seddon, T. (2008). Drugs, the informal economy and globalization. *International Journal of Social Economics, 35*(10), 717-728. doi:http://dx.doi.org/10.1108/03068290810898945

SIC 9532: Administration of Urban Planning and Community and Rural Development. (2011). In L. M. Pearce (Ed.), *Encyclopedia of American Industries* (6th ed., Vol. 3, pp. 3154-3160). Detroit: Gale. Retrieved from http://go.galegroup.com/ps/i.do?id=GALE%7CCX1930401035&v=2.1&u=lirn410 86&it=r&p=GVRL&sw=w&asid=760b014b0ac10d72810ae4c46e21cbca

siteresources.worldbank.org. (2014). The definitions of poverty. Retrieved from http://siteresources.worldbank.org/INTPOVERTY/Resources/335642-1124115102975/1555199-1124115187705/ch2.pdf

Social Exclusion. (2008). In W. A. Darity, Jr. (Ed.), *International Encyclopedia of the Social Sciences* (2nd ed., Vol. 7, pp. 586-589). Detroit: Macmillan Reference USA. Retrieved from http://go.galegroup.com/ps/i.do?id=GALE%7CCX3045302491&v=2.1&u=lirn410 86&it=r&p=GVRL&sw=w&asid=6bfee941d05a3a66f44ba110de7c466e

Stagnation. (2008). In W. A. Darity, Jr. (Ed.), *International Encyclopedia of the Social Sciences* (2nd ed., Vol. 8, pp. 83-85). Detroit: Macmillan Reference USA. Retrieved from http://go.galegroup.com/ps/i.do?id=GALE%7CCX3045302583&v=2.1&u=lirn410 86&it=r&p=GVRL&sw=w&asid=862c567f938f74d50773307ad9ab199d

T, M. A., & Grau, H. R. (2004). Globalization, migration, and latin american ecosystems. *Science, 305*(5692), 1915-1916. Retrieved from http://search.proquest.com/docview/213581385?accountid=160851

Tilahun, M., & Ayele, G. (2013). Factors associated with age at first sexual initiation among youths in gamo gofa, south west ethiopia: A cross sectional study. *BMC Public Health, 13*, 622. doi:http://dx.doi.org/10.1186/1471-2458-13-622

Træen, B., Holmen, K., & Stigum, H. (2007). Extradyadic sexual relationships in norway. *Archives of Sexual Behavior, 36*(1), 55-65. doi:http://dx.doi.org/10.1007/s10508-006-9080-0

twoandahalfmiles.org. (2014). Trafficking Stats. Retrieved from http://twoandahalfmiles.org/trafficking-statistics/

Valera, R. J., Sawyer, R. G., & Schiraldi, G. R. (2000). Violence and post traumatic stress disorder in a sample of inner city street prostitutes. *American*

Journal of Health Studies, 16(3), 149-155. Retrieved from http://search.proquest.com/docview/210478613?accountid=160851

Veenstra, G. (2002). Income inequality and health: Coastal communities in british columbia, canada. *Canadian Journal of Public Health, 93*(5), 374-9. Retrieved from http://search.proquest.com/docview/232004378?accountid=160851

Wabel, N. T. (2011). Psychopharmacological aspects of catha edulis (khat) and consequences of long term use: A review. *Journal of Mood Disorders, 1*(4), 187. doi:http://dx.doi.org/10.5455/jmood.20111217095840

web.worldbank.org. (2011). Choosing and Estimating a Poverty Line. Retrieved from http://web.worldbank.org/WBSITE/EXTERNAL/TOPICS/EXTPOVERTY/EXTPA/0,,contentMDK:20242879~menuPK:435055~pagePK:148956~piPK:216618~theSitePK:430367,00.html

Werlin, H. (1999). The slum upgrading myth. *Urban Studies, 36*(9), 1523-1534. Retrieved from http://search.proquest.com/docview/236262259?accountid=160851

Williams, K. D., Wesselmann, E. D., & Chen, Z. (2007). Social Exclusion. In R. F. Baumeister & K. D. Vohs (Eds.), *Encyclopedia of Social Psychology* (Vol. 2, pp. 896-897). Thousand Oaks, CA: SAGE Publications. Retrieved from http://go.galegroup.com/ps/i.do?id=GALE%7CCX2661100520&v=2.1&u=lirn41086&it=r&p=GVRL&sw=w&asid=687678cfdc03249ecf68c655b71db995

Wirth, Louis 1897-1952. (2003). In *Contemporary Authors* (Vol. 210, pp. 439-441). Detroit: Gale. Retrieved from http://go.galegroup.com/ps/i.do?id=GALE%7CCX3412400299&v=2.1&u=lirn41086&it=r&p=GVRL&sw=w&asid=d28b416830ab74f5a4515b9611b9e5cf

www.aq.upm.es. (2003). THE CHALLENGE OF SLUMS: GLOBAL REPORT ON HUMAN SETTLEMENTS 2003. Retrieved from http://www.aq.upm.es/habitabilidadbasica/docs/recursos/monografias/the_challenge_of_slums-(2003).pdf

www.ecpat.net. (2007). Global Monitoring Report on the status of action against commercial sexual exploitation of children. Retrieved from http://www.ecpat.net/sites/default/files/Global_Monitoring_Report-ETHIOPIA.pdf

www.enddemandillinois.org. (2011). 10 Facts that Shaped the Ugly Truth Campaign. Retrieved from http://www.enddemandillinois.org/10-facts-shaped-ugly-truth-campaign

www.merriam-webster.com. (2014). Prostitution. Retrieved from http://www.merriam-webster.com/dictionary/prostitution

www.oaklandinstitute.org. (2013). DEVELOPMENT AID TO ETHIOPIA. OVERLOOKING VIOLENCE, MARGINALIZATION, AND POLITICAL REPRESSION. Retrieved from http://www.oaklandinstitute.org/sites/oaklandinstitute.org/files/OI_Brief_Development_Aid_Ethiopia.pdf

www.ruralpovertyportal.org. (2014). Rural Poverty in Ethiopia. Retrieved from http://www.ruralpovertyportal.org/country/home/tags/ethiopia

www.unicef.org. (2014). Stop child porn today. Retrieved from http://www.unicef.org/philippines/8895_9845.html

www.washingtontimes.com. (2013). Iran's Middle Class and Part Time Prostitute. Retrieved from http://www.washingtontimes.com/news/2013/may/16/irans-educated-middle-class-and-part-time-prostitu/?page=all

www.who.int. (2014). Urban Health. Retrieved from http://www.who.int/topics/urban_health/en/

www.worldbank.org. (2014). Ethiopia. Retrieved from http://www.worldbank.org/en/country/ethiopia

www.youtube.com. (2014). Part 1/2 - Piers Morgan Interviews Pornstar Belle Knox Duke. Retrieved from http://www.youtube.com/watch?v=3oscJrlaqSQ

Young, A. M., Boyd, C., & Hubbell, A. (2000). Prostitution, drug use, and coping with psychological distress. *Journal of Drug Issues, 30*(4), 789-800. Retrieved from http://search.proquest.com/docview/208845444?accountid=160851

Ziraba, A. K., Kyobutungi, C., & Zulu, E. M. (2011). Fatal injuries in the slums of nairobi and their risk factors: Results from a matched case-control study. *Journal of Urban Health, 88*, 256-65. doi:http://dx.doi.org/10.1007/s11524-011-9580-7

References

Harcourt, C., & Donovan, B. (2005, June). The many faces of sex work. *Sex Transm Infect, 81*(3), 201-6. Retrieved from http://www.ncbi.nlm.nih.gov/pubmed/15923285

aspe.hhs.gov. (2013). 2013 Poverty guidelines. Retrieved from http://aspe.hhs.gov/poverty/13poverty.cfm

hdr.undp.org. (2013). Human Development Report 2013. The Rise of the South:Human Progress in a Diverse World. Retrieved from http://hdr.undp.org/sites/default/files/Country-Profiles/ETH.pdf

newswire.uark.edu. (2014). Affluent, Educated Women May Choose Sexual Prostitution. Retrieved from http://newswire.uark.edu/articles/16181/affluent-educated-women-may-choose-sexual-prostitution

prostitution.procon.org. (2009). 100 Countries and Their Prostitution Policies. Retrieved from http://prostitution.procon.org/view.resource.php?resourceID=000772#ethiopia6.http://prostitution.procon.org/view.resource.php?resourceID=000772#ethiopia

siteresources.worldbank.org. (2014). The definitions of poverty. Retrieved from http://siteresources.worldbank.org/INTPOVERTY/Resources/335642-1124115102975/1555199-1124115187705/ch2.pdf

twoandahalfmiles.org. (2014). Trafficking Stats. Retrieved from http://twoandahalfmiles.org/trafficking-statistics/

web.worldbank.org. (2011). Choosing and Estimating a Poverty Line. Retrieved from http://web.worldbank.org/WBSITE/EXTERNAL/TOPICS/EXTPOVERTY/EXTPA/0,,contentMDK:20242879~menuPK:435055~pagePK:148956~piPK:216618~theSitePK:430367,00.html

www.ecpat.net. (2007). Global Monitoring Report on the status of action against commercial sexual exploitation of children. Retrieved from http://www.ecpat.net/sites/default/files/Global_Monitoring_Report-ETHIOPIA.pdf

www.enddemandillinois.org. (2011). 10 Facts that Shaped the Ugly Truth Campaign. Retrieved from http://www.enddemandillinois.org/10-facts-shaped-ugly-truth-campaign

www.merriam-webster.com. (2014). Prostitution. Retrieved from http://www.merriam-webster.com/dictionary/prostitution

www.oaklandinstitute.org. (2013). DEVELOPMENT AID TO ETHIOPIA. OVERLOOKING VIOLENCE, MARGINALIZATION, AND POLITICAL REPRESSION. Retrieved from http://www.oaklandinstitute.org/sites/oaklandinstitute.org/files/OI_Brief_Development_Aid_Ethiopia.pdf

www.ruralpovertyportal.org. (2014). Rural Poverty in Ethiopia. Retrieved from http://www.ruralpovertyportal.org/country/home/tags/ethiopia

www.washingtontimes.com. (2013). Iran's Middle Class and Part Time Prostitute. Retrieved from http://www.washingtontimes.com/news/2013/may/16/irans-educated-middle-class-and-part-time-prostitu/?page=all

www.worldbank.org. (2014). Ethiopia. Retrieved from http://www.worldbank.org/en/country/ethiopia

www.youtube.com. (2014). Part 1/2 - Piers Morgan Interviews Pornstar Belle Knox Duke. Retrieved from http://www.youtube.com/watch?v=3oscJrlaqSQ

I want morebooks!

Buy your books fast and straightforward online - at one of the world's fastest growing online book stores! Environmentally sound due to Print-on-Demand technologies.

Buy your books online at
www.get-morebooks.com

Kaufen Sie Ihre Bücher schnell und unkompliziert online – auf einer der am schnellsten wachsenden Buchhandelsplattformen weltweit!
Dank Print-On-Demand umwelt- und ressourcenschonend produziert.

Bücher schneller online kaufen
www.morebooks.de

OmniScriptum Marketing DEU GmbH
Heinrich-Böcking-Str. 6-8
D - 66121 Saarbrücken
Telefax: +49 681 93 81 567-9

info@omniscriptum.com
www.omniscriptum.com

www.ingramcontent.com/pod-product-compliance
Lightning Source LLC
Chambersburg PA
CBHW031536210526
45464CB00003B/1028